Soft Computing Techniques in Engineering, Health, Mathematical and Social Sciences

Edge AI in Future Computing

Series Editors:

Arun Kumar Sangaiah
SCOPE, VIT University, Tamil Nadu
Mamta Mittal
G. B. Pant Government Engineering College, Okhla, New Delhi

Soft Computing Techniques in Engineering, Health, Mathematical and Social Sciences
Pradip Debnath and S. A. Mohiuddine

For more information about this series, please visit: https://www.routledge.com/Edge-AI-in-Future-Computing/book-series/EAIFC

Soft Computing Techniques in Engineering, Health, Mathematical and Social Sciences

Edited by
Pradip Debnath
Department of Applied Science and Humanities
Assam University Silchar, India

S. A. Mohiuddine
Department of General Required Courses, Mathematics
Faculty of Applied Studies
King Abdulaziz University
Jeddah, Saudi Arabia

Operator Theory and Applications Research Group
Department of Mathematics, Faculty of Science,
King Abdulaziz University
Jeddah, Saudi Arabia

CRC Press
Taylor & Francis Group
Boca Raton London New York

CRC Press is an imprint of the
Taylor & Francis Group, an **informa** business

First edition published 2021
by CRC Press
6000 Broken Sound Parkway NW, Suite 300, Boca Raton, FL 33487-2742

and by CRC Press

2 Park Square, Milton Park, Abingdon, Oxon, OX14 4RN

CRC Press is an imprint of Taylor & Francis Group, LLC

Library of Congress Cataloging-in-Publication Data
Names: Debnath, Pradip, editor. | Mohiuddine, S. A., editor.
Title: Soft computing techniques in engineering, health, mathematical and social sciences / edited by Pradip Debnath and S.A. Mohiuddine.
Description: Boca Raton, FL : CRC Press, 2021. | Series: Edge AI in future computing | Includes bibliographical references and index. | Summary: "The book contains theory and applications of soft computing in engineering, health, social and applied sciences. Different soft computing techniques such as artificial neural networks, fuzzy systems, evolutionary algorithms and hybrid systems are discussed. It also contains important chapters in machine learning and clustering"— Provided by publisher.
Identifiers: LCCN 2021004280 (print) | LCCN 2021004281 (ebook) | ISBN 9780367750688 (hardback) | ISBN 9780367752545 (paperback) | ISBN 9781003161707 (ebook)
Subjects: LCSH: Soft computing.
Classification: LCC QA76.9.S63 S6385 2021 (print) | LCC QA76.9.S63 (ebook) | DDC 006.3—dc23
LC record available at https://lccn.loc.gov/2021004280
LC ebook record available at https://lccn.loc.gov/2021004281

ISBN: 978-0-367-75068-8 (hbk)
ISBN: 978-0-367-75254-5 (pbk)
ISBN: 978-1-003-16170-7 (ebk)

Typeset in Times
by codeMantra

Contents

Preface

This book is meant for graduate students, faculties and researchers willing to expand their knowledge in any branch of soft computing. The readers of this book will require minimum pre-requisites of undergraduate studies in computation and mathematics. Beginners as well as advanced researchers will find several useful tools and techniques to develop their skills and expertise. The most attractive aspect of the book is that any learner, teacher or researcher from science, engineering or social sciences using soft computing techniques will find this book beneficial. The book contains sufficient theory and applications of soft computing in all branches as suggested by its title. The book presents a survey of the existing knowledge and also the current state-of-the-art development through original new contributions from the researchers. This book may be used as a reference book for a broad range of readers worldwide interested in soft computing. Soft computing consists of a diverse collection of topics. The main sources of information of these topics are scattered among a variety of journals and proceedings. As such, consulting all the information by an amateur learner is seldom possible. Our book is an attempt in this direction of providing a simple interface for an amateur as well as advanced learners in soft computing.

Editors

Pradip Debnath is an assistant professor (in mathematics) in the Department of Applied Science and Humanities of Assam University, Silchar (a central university), India. His research interests include Fuzzy Logic, Fuzzy Graphs, Fuzzy Decision Making, Soft Computing and Fixed-Point Theory. He has published over 50 papers in various journals of international repute. He is a reviewer for more than 20 international journals including Elsevier, Springer, IOS Press, Taylor and Francis and Wiley. He has successfully guided Ph.D. students in the areas of Fuzzy Logic, Soft Computing and Fixed-Point Theory. At present, he is working on a major Basic Science Research Project funded by UGC, Government of India. He received a gold medal in his postgraduation from Assam University, Silchar and qualified several national level examinations in mathematics.

S. A. Mohiuddine is a full professor of mathematics at King Abdulaziz University, Jeddah, Saudi Arabia. As an active researcher, he has coauthored two books, Convergence Methods for Double Sequences and Applications (Springer, 2014) and Advances in Summability and Approximation Theory (Springer, 2018), and published over 140 research papers to various leading journals. He is the referee of many scientific journals and member of the editorial board of various scientific journals, international scientific bodies and organizing committees. He has visited several international universities including Imperial College London, UK. He was a guest editor of a number of special issues for Abstract and Applied Analysis, Journal of Function Spaces and The Scientific World Journal. His research interests are in the fields of sequence spaces, statistical convergence, matrix transformation, measures of noncompactness and approximation theory. His name was in the list of World's Top 2% Scientists (2020) prepared by Stanford University, California.

Contributors

Antonio Abreu
ISEL, Instituto Politécnico de Lisboa
Lisbon, Portugal
and
ISEL-IPL/CTS
Nova Univ of Lisbon
Lisbon, Portugal

Mobeen Ahmad
Department of Mathematics
Aligarh Muslim University
Aligarh, India

P. Balaji
Department of Mathematics
MEASI Academy of Architecture
Chennai, India

Anuradha Bhattacharjee
Department of Mass Communication and Journalism
Assam University
Silchar, India

J. M. F. Calado
IDMEC/ISEL - Instituto Superior de Engenharia de Lisboa
Instituto Politécnico de Lisboa
Lisbon, Portugal

Purnendu Das
Department of Computer Science
Assam University
Silchar, India

Pritam Deb
Department of Physics
Tezpur University
Tezpur, India

Pradip Debnath
Department of Applied Science and Humanities
Assam University
Silchar, India

Kaushik Dehingia
Department of Mathematics
Gauhati University
Guwahati, India

Nitesh Dhiman
Department of Mathematics
C.C.S. University
Meerut, India

Ana Dias
UNINOVA/ISEL - Instituto Superior de Engenharia de Lisboa
Instituto Politécnico de Lisboa
Lisbon, Portugal

Kamini
Department of Mathematics
C.C.S. University
Meerut, India

Vakeel Ahmad Khan
Department of Mathematics
Aligarh Muslim University
Aligarh, India

Omer Kişi
Department of Mathematics
Bartın University
Bartın, Turkey

Korobi Konwar
Department of Physics
Tezpur University
Tezpur, India

K. Loganathan
Department of Mathematics
Erode Arts & Science College
Erode, India

Fernanda Mendes
ESAI-Escola Superior de Atividades Imobiliárias
Lisbon, Portugal

Vishnu Narayan Mishra
Department of Mathematics
Indira Gandhi National Tribal University
Lalpur, India

Bhagya Jyoti Nath
Department of Mathematics
Barnagar College
Sorbhog, India

Satyabrata Nath
Department of Computer Science
Assam University
Silchar, India

Kuldip Raj
School of Mathematics
Shri Mata Vaishno Devi University
Katra, India

N. Revathi
Department of Computer Science
Periyar University PG Extension Centre
Dharmapuri, India

Kavita Saini
School of Mathematics
Shri Mata Vaishno Devi University
Katra, India

Ricardo Santos
GOVCOPP - University of Aveiro, Portugal/ISEL
Instituto Politécnico de Lisboa
Lisbon, Portugal

M. K. Sharma
Department of Mathematics
C.C.S. University
Meerut, India

S. Sivaramakrishnan
Department of Mathematics
Manakula Vinayagar Institute of Technology
Kalitheerthal Kuppam, India

Jose Miguel Soares
ISEG-Lisbon School of Economics & Management
Universidade de Lisboa
Portugal/ADVANCE, ISEG, UL
Lisbon, Portugal

K. Tamilvanan
Department of Mathematics
Government Arts College for Men
Krishnagiri, India

Srinivasan Vijayabalaji
Department of Mathematics
University College of Engineering, Panruti (A Constituent College of Anna University)
Panruti, India

1 Revisiting the Machine Learning Algorithms and Applications in Engineering and Computer Science

Satyabrata Nath, Purnendu Das, and Pradip Debnath
Assam University Silchar

CONTENTS

1.1 INTRODUCTION

Artificial Intelligence (AI) is an intelligence mechanism for machines which enables them to imitate human intelligence and mimic their behavior to a certain extent or better (Charniak 1985, Genesereth and Nilsson 2012). AI was developed keeping in mind to solve complex real-world problems through an approximation of human decision-making capabilities and perform tasks in ever more humane ways (Russell and Norvig 2002).

Machine learning is a subset of Artificial Intelligence that enables the system to learn from past experiences and adapt itself to improve without programming explicitly for each task (Mohri et al. 2018, Alpaydin 2020). This adaptability enables machine learning algorithms to perform operations like prediction or classification based on the available data as well as making them feasible and cost-effective than manual programming (Domingos 2012). Machine learning became more popular with the rise of the internet and the accumulation of abundant data in the form of digital information, which made the engineers realize that instead of teaching the computer/machines to perform every task, it would be more efficient if they connect the system to the internet where all the data is accessible, then it would be possible for the system to learn by itself (Jordan and Mitchell 2015). Due to this automation of systems, machine learning is widely used in various fields consisting of data mining, robotics, image and speech recognition, medical diagnosis, natural language processing, sentiment analysis, classification and prediction, fraud detection and many more.

There are generally three stages for conducting machine learning (Mitchell 1997) – The first stage is called Training Phase where the model is trained with a training dataset in which the inputs are mapped with expected outputs. This phase prepares the machine learning model for classification and prediction tasks. Stage two is the validation and testing phase where we test the model with some test samples and measure its performance that how well it has been trained. Measures like errors, accuracy, time and space complexities, etc. are observed. The third and final stage is the application stage where the model is introduced to the real world to extract necessary information and solve real-life problems.

If we look back into the past of machine learning origins, Alan Turing in the year 1950 developed the "Turing Test" to examine if a computer has human-like intelligence (Turing 1950). In the 1950s, Arthur Samuel created a computer program for playing checkers and made the term "machine learning" popular (Samuel 1959). Rosenblat invented "Perceptron" in 1957, an electronic device which worked on the principles of biological learning mechanism and simulated human thought process (Rosenblatt 1957). In 1967, the algorithm "Nearest Neighbor" was developed, enabling computers to start using very simple pattern recognition. This could be used to map a path for traveling salesmen who began at a random city but ensuring that they visit all cities within a short tour (Cover and Hart 1967). Kunihiko Fukushima introduced "Neocognitron" in 1979 which was a multilayered ANN and was used for the identification of handwritten characters and other tasks of pattern recognition and laid the groundwork for convolutional neural networks (Fukushima 1980). In 1982, John Hopfield popularized Hopfield networks, which could act as a content addressable network for recurring neural network applications (Hopfield 1982). In 1986, the backpropagation method was popularized for ANNs (Rumelhart et al. 1986), and in 1989, Christopher Watkins developed Q-learning, which dramatically increases the practicality and effectiveness of reinforcement learning (Watkins 1989). In 1995, two popular methods were introduced in the field of machine learning. A paper explaining random forest decisions was published by Tin Kam Ho (Ho 1995), whereas Corinna Cortes and Vladimir Vapnik have presented their research on Support Vector Machines (Cortes and Vapnik 1995). After that, many methods and algorithms were developed in this arena and are still advancing with the provision of accessing large data from everywhere, thereby efficiently utilizing machine learning in every aspect.

1.2 DIFFERENT LEARNING APPROACHES USED BY MACHINE LEARNING

Machine learning uses many types of learning techniques which basically fall under the following as shown in Figure 1.1

a. **Supervised Learning** – In this type of learning, the machine learns under supervision. It is first provided with a set of labeled data and training data (Kotsiantis et al. 2006). Here, the training data serve as an instructor with inputs coupled with the correct outputs (Mohri et al. 2018). The learning algorithm in its training phase will search for certain patterns and features to relate with the labeled outputs. Following the training phase, the supervised learning algorithm will be assigned new inputs and evaluate the label of new inputs based on prior training results.
b. **Unsupervised Learning** – Here, the machine learns on its own without any guidance. The learning algorithm, in this case, is not provided with any labeled data leaving it to find hidden patterns among the input data on its own in order to predict the output (Ghahramani 2003). Whenever a new input is introduced, the algorithm classifies it based upon the previous data it has been trained on.
c. **Semi-supervised Learning** – Semi-supervised learning is in the intermediate state between supervised and unsupervised learning (Zhu and Goldberg 2009). Supervised learning uses labeled data, which is very expensive, whereas unsupervised learning uses unlabeled data. For maximizing prediction, semi-supervised learning is utilized in cases where there are minimum labeled data and maximum unlabeled data.
d. **Reinforcement Learning** – This type of learning is based upon an agent (the learning algorithm) and its interaction with a dynamic environment (Sutton 1992). The interaction here is mainly trial and error where the agent at a particular state performs an action based on a situation and it gets rewarded for a correct decision and penalized for an incorrect one. Based upon the positive rewards, the model trains itself and becomes ready to predict new data when presented.

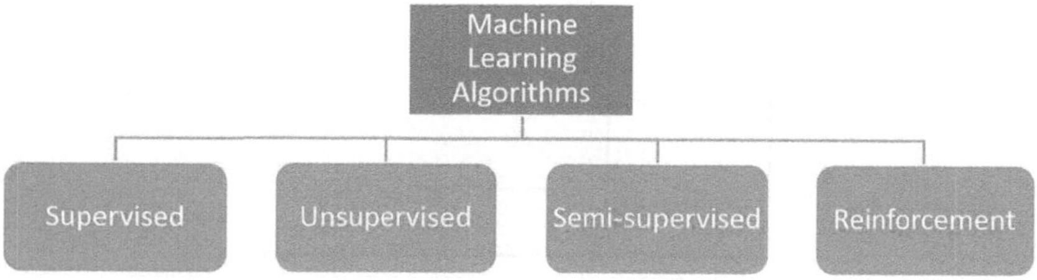

FIGURE 1.1 Representational model of machine learning approaches.

1.3 ALGORITHMS USED BY DIFFERENT LEARNING APPROACHES

1.3.1 Supervised Learning Algorithms

Supervised algorithms can be further categorized into algorithms for regression and classification. Classification algorithms are used for predicting distinct class labels, whereas regression algorithms are applied to predict continuous data (Caruana 2006).

A. **Linear Regression** – This statistical method is used to predict the dependent variable Y with respect to values of independent variables X. Linear regression is used for predictive analysis of continuous quantity (Ray 2019, Montgomery et al. 2012). Mathematically, it can be represented as follows –

$$Y = f_0 + f_1X + e \tag{1.1}$$

where Y is the dependent variable, f_0 is the Y intercept, f_1 is the slope, X is the independent variable and e is some random error. Figure 1.2 represents the structure of linear regression.

B. **Logistic Regression** – Logistic regression falls under the classification algorithm which is used to predict dependent variable Y for given independent variables X, where the dependent variables are categorical. This technique uses a cost function called sigmoid function instead of a linear function and is majorly used for binary classification where we need to predict the probability of the dependent variable that it belongs to which class (Dreiseitl and Machado 2002).

C. **Decision Trees** – This approach is one of the most prominent and extensively used techniques that can be used for both classification where prediction of output is categorical value and regression problems where prediction of output is continuous value (Kotsiantis 2013). A decision tree structure consists of nodes and branches where the internal nodes denote the attribute

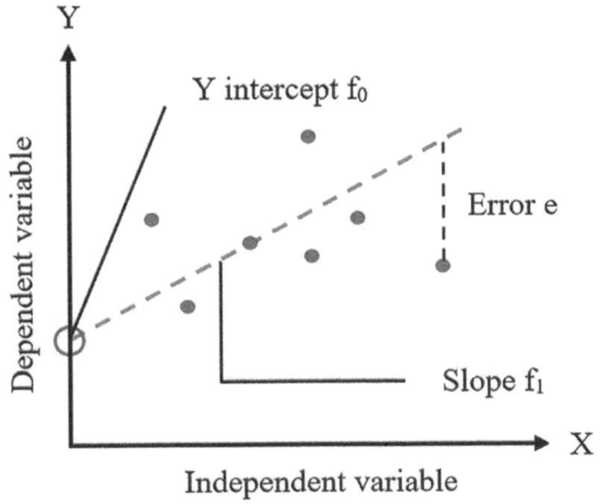

FIGURE 1.2 Linear regression description.

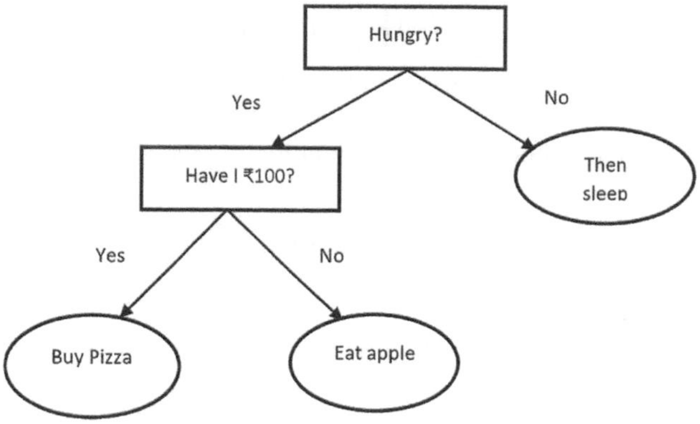

FIGURE 1.3 A graphical illustration of a simple decision tree.

conditions, the class labels are represented by the leaf/terminal node, and the branches denote the value to be considered (Kotsiantis et al. 2007). An example of a decision tree structure is illustrated in Figure 1.3.

D. **Naïve Bayesian Model** – Naïve Bayes algorithm can be applied for both discrete and continuous attributes, but mostly it is used for classification problems. This method is famous because it is easy to construct and interpret which makes it suitable for large datasets (Wu et al. 2008). Naïve Bayes is based upon the fundamentals of Bayes theorem that operates on conditional probability. The conditional probability basically means the probability of occurrence of an event given that the probability of some other event has already occurred (Gut 2013). Hence, the Bayes theorem can be described as follows –

$$P(Y \mid X) = \frac{P(X \mid Y)P(Y)}{P(X)} \tag{1.2}$$

where X and Y are two events that are considered. $P(Y|X)$ is the posterior probability of Y given X has occurred, $P(Y)$ is the prior probability of Y, $P(X)$ is the marginal probability of X and $P(X|Y)$ is the likelihood probability of X given Y has occurred. **Naïve Bayesian classifier** (Rish 2001) is an application of Naïve Bayes algorithm that utilizes the Bayes theorem for classification problems. Here, Y may be any particular record that is to be categorized depending upon X, where $X = (x_1, x_2 \ldots x_n)$ is a set of independent features. So, equation (1.2) can be rewritten as follows –

$$P(Y \mid (x_1, x_2 \ldots x_n)) = \frac{P(x_1 \mid Y)P(x_2 \mid Y)\ldots P(x_n \mid Y)P(Y)}{P(x_1)P(x_2)\ldots P(x_n)} \tag{1.3}$$

$$= \frac{P(Y)\prod_{i=1}^{n} P(x_1 \mid Y)P(x_2 \mid Y)\ldots P(x_n \mid Y)}{P(x_1)P(x_2)\ldots P(x_n)} \tag{1.4}$$

$$= \frac{P(Y)\prod_{i=1}^{n} P(x_i \mid Y)}{P(x_1)P(x_2)\ldots P(x_n)} \tag{1.5}$$

Since the denominator remains constant for all inputs, it can be removed

$$P\big(Y \mid (x_1, x_2 \ldots\ldots x_n)\big) \propto P(Y)\prod_{i=1}^{n} P(x_i \mid Y) \tag{1.6}$$

The class variable Y with the highest likelihood must be calculated using all results of the class variable

$$Y = \text{argmax}_Y \; P(Y)\prod_{i=1}^{n} P(x_i \mid Y) \tag{1.7}$$

E. **Support Vector Machine** – In order to resolve the classification and regression task, Vapnik introduced the Support Vector Machine (SVM) that produces precision results even with a limited quantity of data regardless of dimensionality factor (Cortes and Vapnik 1995). Constructing a hyperplane for separating groups of data points is the working theory of SVM. It also draws two marginal lines along with the hyperplane in such a way that the distance between the two marginal lines is maximum, thereby maximizing the precision of the classification, to ensure that the problem is readily linearly separable. The SVM algorithm is popular as it can perform not only linear classification but also non-linear classification through kernel trick (Theodoridis 2008). The main goal of SVM kernels is to transform lower dimensions into higher dimensions so that the hyperplanes can easily categorize non-linear data points. Figure 1.4 graphically demonstrates the SVM.
F. **K-Nearest Neighbors** – K-Nearest Neighbors (KNN) is considered to be the most basic yet fundamental supervised learning algorithm for classification purposes. This non-parametric technique is a form of lazy learning (Bontempi et al. 1999), where the function is only localized, and all calculations are postponed before function analysis is carried out. When a test sample is introduced for the classification process, the algorithm starts searching all the instances in the training set by finding the distances of the test sample with all the respective training data to determine the nearest neighbors of the test sample and classify it according to the maximum nearest neighbors with the shortest distance (Mitchell 1996).
G. **Artificial Neural Networks** – Artificial Neural Networks (ANN) is a machine-learning technology motivated by the human brain's neural network system. ANN consists of a series of interconnected nodes that represent artificial neurons, similar to the neurons' connections in our brain (Yegnanarayana 2009). These ties between the nodes are like brain synapses, which allow the transmission of signals from one node to another.

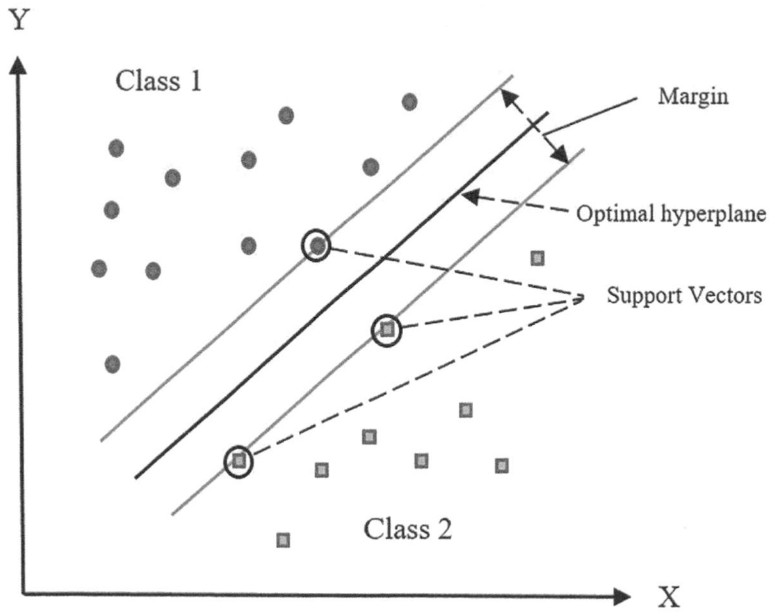

FIGURE 1.4 Pictorial representation of support vector machine classifying two classes.

Inputs in the form of real numbers are processed in the nodes, and the sum of inputs is determined by a non-linear activation function. Each node is linked to other nodes via edges with certain weights. The ANN learns by changing the weight values (Priddy and Keller 2005). The ANN comprises three classes of nodes: the input layer, hidden layer and the output layer (Bre et al. 2018) as shown in Figure 1.5.

In the supervised version of ANN, the output is already known, and the predicted output is matched with the actual output (Ojha et al. 2017). The parameters will be modified depending on the error which will be then back propagated to adjust the weights and the process will continue till the predicted outputs match the expected output.

1.3.2 Unsupervised Learning Algorithms

Two main algorithms are majorly used in unsupervised learning, which are as follows–

A. **K Means Clustering** – The K Means algorithm is a partitioning clustering algorithm which partitions the unlabeled datasets into K clusters by finding the optimal centroid in the high dimensional vector space (Kanungo et al. 2002). In this iterative process, the data points are allocated to a specific cluster by calculating the minimum distance (Euclidean or Manhattan)

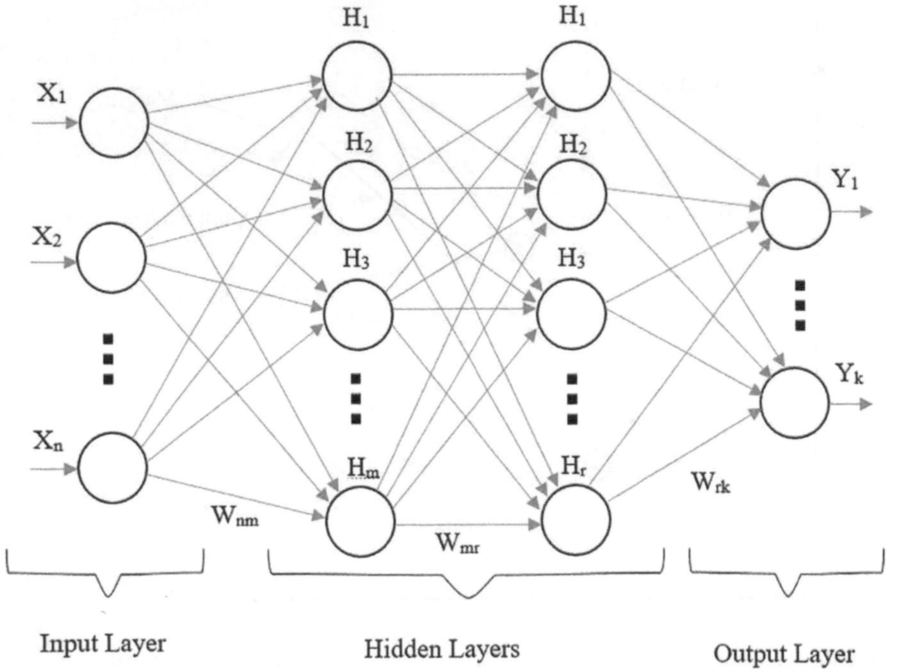

FIGURE 1.5 A portrayal of multi-layered artificial neural network.

from the data points to the centroids (Kapil and Chawla 2016). The process of clustering is represented in Figure 1.6.

B. **Principal Component Analysis** – Primary Component Analysis (PCA) is a dimension reduction statistical approach frequently used to decrease the dimensionality of large datasets by converting a large set to smaller variables that still have the majority of information within that large set (Wold et al. 1987). Karl Pearson invented this technique in the year 1901

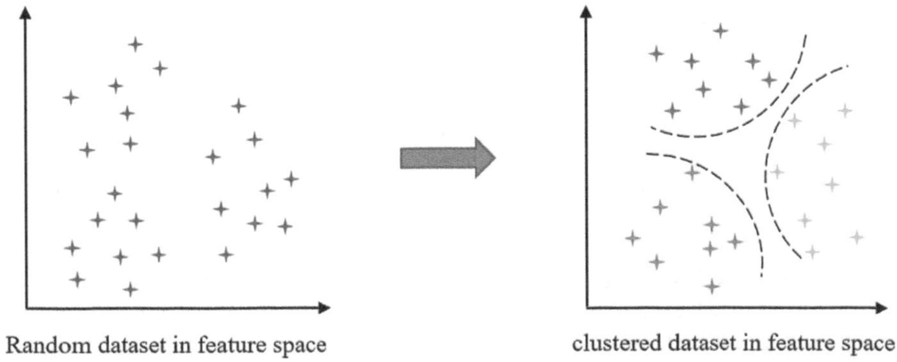

FIGURE 1.6 K-Means clustering procedure.

(Pearson 1901), and it was further developed by Harold Hotelling in 1933 (Hotelling 1933). This machine-learning clustering algorithm can be applied for noise filtration, image compression and many more tasks.

C. **Apriori Algorithm** – A breadth-first search method known as the apriori algorithm (Agarwal and Srikant 1994) calculates the support between elements. This support effectively maps one data item's reliance on another, which can help us to understand which data item affects the probability of something happening to the other data item (Yu et al. 2008). Bread, for instance, influences the customer to purchase milk and eggs. So, mapping helps maximize the store's profits. Using this algorithm, which yields rules for its output, mapping of this kind can be learned.

1.3.3 Semi-supervised Learning Algorithms

A. **Self-Training** – Self-Training is a self-taught algorithm which is generally used in semi-supervised learning (Zhu and Goldberg 2009). In this process, a classifier is trained first with the limited amount of data which are labeled. After training, the classifier is then used for the unlabeled data classification. The training set normally contains the most accurate unlabeled points along with their labels predicted (Zhu 2005). This process is repeated till an optimal point. In order to learn itself, the classifier uses its own predictions and thus is called the self-training method.

B. **Graph-Based** – This learning method constructs a graph $G = (V, E)$ from the training data where V represents the vertices that are labeled and unlabeled data, and E denotes the undirected edges connecting the nodes i, j with weights $w_{i,j}$. The similarity of the two instances is expressed by an edge between two vertices x_i, x_j (Zha et al. 2009). Figure 1.7 represents a graph structure where there are some labeled nodes and the majority of unlabeled nodes. All nodes are connected via edges. Weights play an important part

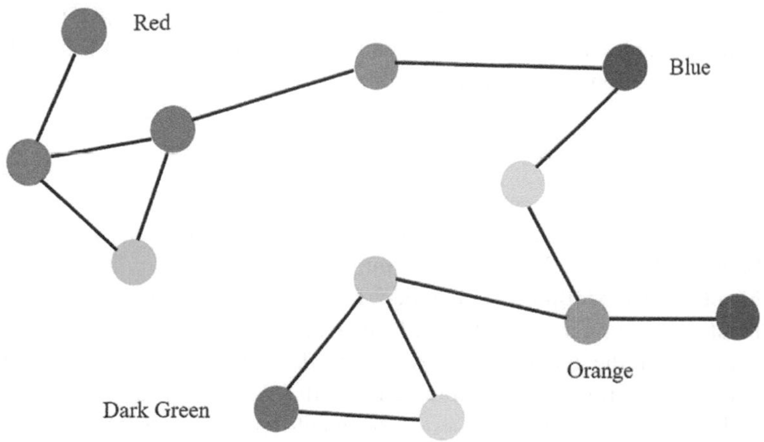

FIGURE 1.7 Depiction of graph-based semi-supervised learning.

here, because if weight is greater, then two labels y_i, y_j are likely to be similar (Zhu and Goldberg 2009). The edge weights are often defined by certain weight estimating functions some of which are –Gaussian edge weight function which is defined as $w_{ij} = e\left(-\left\|x_i - x_j\right\|^2 / \sigma^2\right)$, KNN edge weight function defined as $w_{ij} = 1$ if x_i is within the proximity of k-nearest neighbors of x_j or same way around, and $w_{ij} = 0$ otherwise, and many more like Є-radius neighbors, b-matching, etc. (Zhu 2005).

Large w_{ij} means that the $f(x_i)$ and $f(x_j)$ predictions are preferred. It can be formulated by a function f of graph energy as follows –

$$\sum_{i,j=1}^{l+u} w_{ij}\left(f(x_i) - f(x_j)\right)^2 \tag{1.8}$$

C. **Semi-Supervised Generative Adversarial Networks** – The semi-supervised GAN (SGAN) is a generative adversarial network, which has a multi-class classification as a discriminator. SGAN learns to distinguish between $N+1$ classes instead of differentiating between two classes (real and false), where N is the number of classes in the training dataset with an additional one for the fake examples generated by the Generator (Odena 2016, Springenberg 2015). The training process in SGAN is done simultaneously in supervised and unsupervised modes.

1.3.4 Reinforcement Learning Algorithms

A. **Q Learning** – In Q learning, a reinforcement learning agent tries to find out the quality of its behavior based on the rewards it receives so that it can determine which actions to perform in sequence to achieve full long-term rewards (Watkins and Dayan 1992). As it learns evaluating the optimal policy regardless of the agent's action, it is also known as an off-policy reinforcement learner. The main component of this algorithm is a Bellman equation (Sutton 2018) which is defined by

$$Q(S_t, A_t) \leftarrow Q(S_t, A_t) + \alpha\left[R_{t+1} + \gamma \max_a Q(S_{t+1}, a) - Q(S_t, A_t)\right] \tag{1.9}$$

where R is the reward for state transition $S_t \rightarrow S_{t+1}$, α is the learning rate between 0 and 1 and γ is the discount factor.

B. **Monte Carlo Method** – This model-free (Kaelbling et al. 1996) reinforcement learning method can be used where an environment model is provided about which the agent is unaware of, and the only way to gain information about the environment is by interacting with it (Andrieu et al. 2003, Watkins and Dayan 1992). In this process, the agent produces experienced samples and then calculates the value for a state or state action based on average return.

1.4 ANALYSIS OF THE LEARNING ALGORITHMS

From the previous section, we got to know the various learning algorithms with their formal definition, respective to each learning method. Based upon the learning strategy, these algorithms help in classification and prediction. In Table 1.1, a comparative analysis is represented between the various learning strategies, and in Table 1.2, applications of the learning algorithms along with their merits and demerits are illustrated.

1.5 RELATED WORKS IN THE FIELD OF ENGINEERING AND COMPUTER SCIENCE

In this section, some related works and applications of machine learning algorithms are discussed. Jhalani and the team (Jhalani et al. 2016) applied a multi-linear regression approach to incorporate various ratings in collaborative filtering to increase the overall accuracy of the recommender system. For a better computational process, the computer must have good and efficient processors. Joseph et al. 2006 used simulations based on linear regression to assist processor architects to make critical design decisions. A novel text categorization model was proposed by (Genkin et al. 2007) based on the Bayesian logistic regression approach that used Laplace prior to reduce overfitting issues.

One of the important applications of machine learning techniques is the recognition of objects by studying certain features from data. Nasien et al. 2010 used SVM to recognize English handwritten characters. The features of the image data were generated by Freeman Chain Code which was then classified by SVM to obtain greater accuracy. Zhang et al. 2006 proposed a hybrid method (SVM + K-NN) for visual

TABLE 1.1
Comparative Analysis of Machine Learning Strategies

Comparison Criteria	Supervised	Unsupervised	Semi-supervised	Reinforcement
Suitable for	Classification and Regression	Clustering and Association	Classification and Regression	Reward-Based Decision-Making
Training	Under external supervision	No external supervision	Pseudo - labeling or self-training	No external supervision
Data Types	Labeled	Unlabeled	Both labeled & unlabeled	Based upon environment interaction
Methodology	Mapping of labeled inputs to familiar outputs	Finds patterns to predict output	Labels unlabeled data with labeled data for prediction	Makes decision by interacting with the environment
Applications	Sentiment analysis, image classifications, Spam filtration	Data reduction, anomaly detection	Speech analysis, web content classification, protein sequence classification	Robotics, self-driving cars, gaming

TABLE 1.2
Illustration of Machine Learning Algorithm's Merits-Demerits Along with Their Applications

Learning Algorithm	Merits	Limitations	Applicability
Linear regression	• Easy implementation & interpretation • High performance with linearly separable datasets	• Vulnerable to noise & over-fitting • Susceptible to outliers • Prediction of numeric input only	• Forecasting risk and opportunities in business • Trend line representation • Engine performance Analyzation
Logistic regression	• Easy implementation & interpretation • Deal with non-linearity • Classify multi-class problems	• Complex relations are difficult to capture • For stable prediction requires large dataset and training time • Sensitive to outliers	• Image segmentation & classification • Geographic image processing • Handwriting recognition
Decision trees	• Normalization of data not required • The tree structure can easily be understood • Missing values doesn't hamper the training process	• Lack of stability • High time and space complexity • High chance of overfitting. • Calculations become complex if values are uncertain	• Remote sensing • Stock trading • Health care operations • Customer-relationship management
Naïve Bayes	• Less training time • Easy implementation • Insensitive to irrelevant traits	• Assumes all features are independent • Suffers "zero frequency" problem • Performance reduces with more data	• Text classification • Recommendation system • Credit scoring • Real-time prediction • Classification of medical data
SVM	• Supports linear & non-linear classification • Suitable for high dimensional data • Greater accuracy • Can work with unstructured & semi-structured data	• Selection of optimal kernels can be hard • Slow testing phase • Not favorable for large datasets • Less performance in case of overlapping classes	• Face detection • Bioinformatics • Handwriting recognition • Image classification • Speech recognition

(Continued)

TABLE 1.2 (*Continued*)
Illustration of Machine Learning Algorithm's Merits-Demerits Along with Their Applications

Learning Algorithm	Merits	Limitations	Applicability
K-NN	• Training phase not required • Suitable for multi-class problems • Simplistic design	• Not suited to large datasets • Depends on • favorable K selection • Requires scaling of features	• Text mining • Analyzing financial risks • Forecasting of stock market • Blood glucose estimation • Climate forecasting
ANN	• Can learn non-linear and complex relations • Fault-tolerant • Able to work with partial information • Ability to parallel processing	• Requires powerful processing hardware • Slow training • Unexplainable network behavior	• Classification of images • Speech recognition • Face & object detection • Natural language processing • Medical diagnosis • Fingerprint recognition
K Means	• Simple implementation • Adaptable to new data • Scalable to large datasets	• K-value is hard to predict. • Less performance in case of heavily overlapping data • Sensitive to noise & outliers	• Document classification • Fraud detection • Call record analysis • Search details grouping in search engines
PCA	• Removal of correlated features • Enhances the algorithm's performance • Limits overfitting • Improves data visualization	• Becomes less interpretable for independent variables • Data standardization is mandatory • Loss of information	• Image compression • Stocks analyzation • Risk management • Analysis of spike-triggered covariance in neuroscience • Face recognition
Apriori	• Easy understanding & implementation • Proper utilization of large dataset • Easily paralyzed	• Needs several scans of the database • Unfavorable for large sets of data • Slow compared to others	• Finds correlations between people's diabetic condition • Product recommendation system • Market basket analysis

(*Continued*)

TABLE 1.2 (*Continued*)
Illustration of Machine Learning Algorithm's Merits-Demerits Along with Their Applications

Learning Algorithm	Merits	Limitations	Applicability
Self-training SSL	• Simple than other SSL methods • Fast method • Based on Wrapper method	• Sometimes results in noise amplification • Not suitable for discriminative classifiers • Training procedures required at every iteration	• Used in word sense disambiguation • Object detection
Graph-based SSL	• Non-parametric, discriminative and transductive • Better efficiency if the graph suits the job • Fast computation	• Poor performance if the graph does not suit the task • Improper graph structure or edge weights degrades the overall performance	• Handwriting classification • Image segmentation • Face beauty analysis
SGAN	• Good performance with less training samples • Provide efficient predictions • Contains descriptive decision boundaries	• If the amount of unlabeled data is much more compared to label data might result in inconsistency • Efforts to the model are far more demanding than discriminatory models	• Build image dataset examples • Image enhancement • Video prediction
Q learning	• Model-free approach • Reacts to environmental change • Able to find the optimal action-selection policy	• Not suitable for continuous work system • It can only update Q values for those states & actions that it has experienced. • Long computation time	• Automatic configuration of online web systems • Traffic signal controller for networks • Recommendation systems for news
Monte Carlo	• It has zero biasing • Easy to understand • Independent of unrealistic assumptions	• It has to wait until the end of the episode is known. • Contain high variance • Learn only from complete sequences	• Risk management in financial sectors • Wind energy analysis • Computer graphics

category recognition. They tested their model using some popular datasets (MNIST, USPS, CUReT, Caltech-101) and achieved excellent results. Roli and Marcialis 2006 took the help of a self-training approach and incorporated it with the classical PCA technique to build a facial recognition system. It was observed that self-learning can considerably increase the performance even with a minimalistic labeled dataset.

In the field of geotechnical engineering (Goh and Goh 2007), SVM can be used to design a sophisticated relationship between seismic and soil parameters along with liquefaction potential. A novel approach to diagnose marine main engine cylinder cover faults was proposed by Zhan and others (Zhan et al. 2007) where they used SVM to diagnose faults by classifying different types of faults.

Decision trees are one of the popular techniques that are used for classification. Prasad et al. 2015 with the help of decision trees performed sentiment classification of Twitter datasets in the Hindi language. The decision tree is also extensively used in engineering fields. They are used to develop predictive models for developing building energy demand (Yu et al. 2010). In the domain of sound engineering, Amarnath et al. 2013 used decision trees for fault diagnosis in bearings used for industrial machinery by classifying sound signals of good and faulty bearings, whereas Saimurugan et al. 2011 used decision tree and SVM for fault analysis in shafts and bearings.

Naïve Bayes can be used for text classification (Chen et al. 2009) and intrusion detection in networks (Panda and Patra 2007). Muniyandi et al. (2011) applied a cascaded algorithm containing K-Means clustering and C4.5 algorithms for network intrusion detection. ANN is used in Natural Language Processing because of its modularity and adaptability (Martinez et al. 1995). They are also popular for image classification tasks (Ciregan et al. 2012, Cireşan et al. 2012). Their application also extends to process engineering (Willis 1991), rock engineering (Yang and Zhang 1997), geotechnical engineering (Shahin et al. 2001) and many more.

The semi-supervised model PCA is mainly used in the field of image processing. A novel scheme of image denoising by using PCA was presented by Zhang et al. (2010). Du and Fowler proposed a design based on PCA to compress Hyperspectral Image (Du and Fowler 2007).

Yang and his team used neural Q-learning for mobile robot navigation (Yang et al. 2004). A multi-agent Q-learning approach for controlling traffic lights in a non-stationary environment was presented in Abdoos et al. 2011. Taghipour and Kardan proposed a hybrid framework based on Q-learning for web recommendation system (Taghipour and Kardan 2008).

1.6 DISCUSSION

Machine learning algorithms are vastly applicable to numerous real-world problems. They are divided into learning strategies based on human intelligence and try to solve complex problems based on methods belonging to each strategy. Machine learning has a huge number of methods and variations of methods each having its pros and cons, but it is outside our scope to cover all the algorithms and their comparison in this paper. Despite this, we believe that the following assertions will allow the reader to get a better idea of the methods based on their strength, weaknesses and applications.

In supervised learning, artificial neural networks and SVMs can efficiently handle multi-dimensional and continuous data, whereas algorithms based on logics and conditions can very well manage discrete data. SVM, ANN and logistic regression require large data samples to be trained on to provide correct predictions, whereas the Naïve Bayesian model requires fewer training data. K-NN and linear regression cannot properly handle noise, and its similarity index can be easily affected by errors in the attributes which leads to misclassification of the new input by wrong misleading nearest neighbors. Unlike them, logistic regression and most of the decision trees are less affected by noise and overfitting issues because of their pruning methodologies. Generally, K-NN is sensitive to features that are irrelevant which can be understood by the operation of its algorithm; moreover, the structure of the decision tree can be changed because of a small modification in data leading to instability. For ANN, the existence of irrelevant features may render the training phase making it impractical and inefficient. In the case of operating mechanism, linear regression, logistic regression, Naïve Bayes and decision trees are easily interpretable, and their implementation is also rather simple. On the contrary, SVM, K-NN and ANN are poorly interpretable and have complex operations. Although K-NN is not much interpretable, the method is quite transparent and simple. Along with K-NN, Naïve Bayesian models and decision trees are also very much transparent and can be easily understood by users belonging to different domains.

Among various unsupervised learning strategies, we have considered K Means clustering, PCA and apriori algorithm in this study. Implementation of K Means and apriori algorithms are simple and easily understandable compared to PCA, whereas PCA is much more noise-tolerant than the other two methods and prevents overfitting of data. In fact, K Means is very much susceptible to noise and outliers and can easily lose performance in case of overlapping data.

Semi-supervised methods train with the combination of labeled and unlabeled datasets where the number of unlabeled attributes is more. Here three popular semi-supervised methods are reviewed which are Self-Training SSL, Graph-Based SSL and SGAN. Self-Training and Graph-Based algorithms are the fastest in comparison to other SSL methods and the Self-Training method is the simplest among the rest. SGAN can perform better even with less training datasets and provide efficient predictions though the performance can deteriorate if the number of unlabeled samples is much greater than labeled samples. Similarly, in the Graph-based method if the structure of the graph is not appropriate with respect to the task then the performance of the system will be degraded.

For reinforcement learning, two methods are taken into consideration – Q learning and Monte Carlo. The Monte Carlo method has less biasing and high variance compared to Q learning which has the exact opposite feature. Bias tests the contribution of the classifier's central inclination to error when trained on separate data, whereas Variance is a measurement of the contribution of deviations from the central inclination to error (Bauer & Kohavi 1999). Q learning utilizes the Markov property, whereas the Monte Carlo method doesn't.

1.7 CONCLUSION

In this paper, a review of machine learning algorithms along with their advantages, disadvantages and applications is presented. We tried to cover most of the prominent and commonly used techniques in the field of machine learning and compared each technique wherever possible. We reviewed machine learning techniques based on supervised, unsupervised, semi-supervised and reinforcement learning strategies. Examples of related works in the field of engineering and computer science are also discussed. It has been observed that no algorithm is dominant in all the situations and has its own pros and cons and are different from each other based on areas of application. Thus, selecting a suitable algorithm depends upon the nature of the problem and available data. We believe that the readers will get a better understanding of this emerging field from this study and aid them in advancing in this field by providing fascinating research ideas.

REFERENCES

Abdoos, M., N. Mozayani, and A.L.C. Bazzan. 2011. Traffic light control in non-stationary environments based on multi agent Q-learning. In *2011 14th International IEEE Conference on Intelligent Transportation Systems (ITSC)*, pp. 1580–1585. IEEE, Washington DC, New York.

Agarwal, R., and R. Srikant. 1994. Fast algorithms for mining association rules. In *Proceeding of the 20th VLDB Conference*, pp. 487–499, Santiago de Chile.

Alpaydin, E. 2020. *Introduction to Machine Learning*. MIT Press, London, UK.

Amarnath, M., V. Sugumaran, and H. Kumar. 2013. Exploiting sound signals for fault diagnosis of bearings using decision tree. *Measurement* 46(3): 1250–1256.

Andrieu, C., N. D. Freitas, A. Doucet, and M. I. Jordan. 2003. An introduction to MCMC for machine learning. *Machine Learning* 50 (1–2): 5–43.

Bauer, E., and R. Kohavi. 1999. An empirical comparison of voting classification algorithms: Bagging, boosting, and variants. *Machine learning* 36 (1–2): 105–139.

Bontempi, G., M. Birattari, and H. Bersini. 1999. Lazy learning for local modelling and control design. *International Journal of Control* 72 (7–8): 643–658.

Bre, F., J. M. Gimenez, and V. D. Fachinotti. 2018. Prediction of wind pressure coefficients on building surfaces using artificial neural networks. *Energy and Buildings* 158: 1429–1441.

Caruana, R., and A. N. Mizil. 2006. An empirical comparison of supervised learning algorithms. In *Proceedings of the 23rd International Conference on Machine Learning*, pp. 161–168, Pittsburgh.

Charniak, E. 1985. *Introduction to Artificial Intelligence*. Pearson Education India.

Chen, J., H. Huang, S. Tian, and Y. Qu. 2009. Feature selection for text classification with Naïve Bayes. *Expert Systems with Applications* 36 (3): 5432–5435.

Ciregan, D., U. Meier, and J. Schmidhuber. 2012. Multi-column deep neural networks for image classification. In *2012 IEEE Conference on Computer Vision and Pattern Recognition*, pp. 3642–3649. IEEE, Providence, RI, USA.

CireşAn, D., U. Meier, J. Masci, and J. Schmidhuber. 2012. Multi-column deep neural network for traffic sign classification. *Neural Networks* 32: 333–338.

Cortes, C., and V. Vapnik. 1995. Support-vector networks. *Machine Learning* 20 (3): 273–297.

Cover, T., and P. Hart. 1967. Nearest neighbor pattern classification. *IEEE Transactions on Information Theory* 13 (1): 21–27.

Domingos, P. 2012. A few useful things to know about machine learning. *Communications of the ACM* 55 (10): 78–87.

Dreiseitl, S., and L. O. Machado. 2002. Logistic regression and artificial neural network classification models: A methodology review. *Journal of Biomedical Informatics* 35 (5–6): 352–359.

Du, Q., and J. E. Fowler. 2007 Hyperspectral image compression using JPEG2000 and principal component analysis. *IEEE Geoscience and Remote sensing letters* 4 (2): 201–205.

Fukushima, K. 1980. A self-organizing neural network model for a mechanism of pattern recognition unaffected by shift in position. *Biological Cybernetics.* 36 (4): 193–202.

Genesereth, M. R., and N. J. Nilsson. 2012. *Logical Foundations of Artificial Intelligence.* Morgan Kaufmann, Burlington, Massachusetts, USA.

Genkin, A., D. D. Lewis, and D. Madigan. 2007. Large-scale Bayesian logistic regression for text categorization. *Technometrics* 49 (3): 291–304.

Ghahramani, Z. 2003. Unsupervised learning. In *Summer School on Machine Learning*, pp. 72–112. Springer, Berlin, Heidelberg.

Goh, A.T.C., and S.H. Goh. 2007. Support vector machines: Their use in geotechnical engineering as illustrated using seismic liquefaction data. *Computers and Geotechnics* 34(5): 410–421.

Gut, A. 2013. *Probability: A Graduate Course.* Vol. 75. Springer Science & Business Media, Springer-Verlag New York.

Ho, T. K. 1995. Random decision forests. *Proceedings of the Third International Conference on Document Analysis and Recognition*, Montreal, Quebec: IEEE. 1: 278–282.

Hopfield, J. 1982. Neural networks and physical systems with emergent collective computational abilities. *Proceedings of the National Academy of Sciences of the United States of America.* 79 (8): 2554–2558.

Hotelling, H. 1933. Analysis of a complex of statistical variables into principal components. *Journal of Educational Psychology* 24 (6): 417.

Jhalani, T., V. Kant, and P. Dwivedi. 2016. A linear regression approach to multi-criteria recommender system. In *International Conference on Data Mining and Big Data*, pp. 235–243. Springer, Cham,

Jordan, M. I., and T. M. Mitchell. 2015. Machine learning: Trends, perspectives, and prospects. *Science* 349 (6245): 255–260.

Joseph, P. J., K. Vaswani, and M. J. Thazhuthaveetil. 2006. Construction and use of linear regression models for processor performance analysis. In *The Twelfth International Symposium on High-Performance Computer Architecture*, pp. 99–108. IEEE, Austin, Texas.

Kaelbling, L. P., M. L. Littman, and A. W. Moore. 1996. Reinforcement learning: A survey. *Journal of Artificial Intelligence Research* 4: 237–285.

Kapil, S., and M. Chawla. 2016. Performance evaluation of k-means clustering algorithm with various distance metrics. In *2016 IEEE 1st International Conference on Power Electronics, Intelligent Control and Energy Systems (ICPEICES)*, pp. 1–4. IEEE, Delhi, India.

Kanungo, T., D. M. Mount, N. S. Netanyahu, C. D. Piatko, R. Silverman, and A. Y. Wu. 2002. An efficient k-means clustering algorithm: Analysis and implementation. *IEEE Transactions on Pattern Analysis and Machine Intelligence* 24 (7): 881–892.

Kotsiantis, S. B. 2013. Decision trees: A recent overview. *Artificial Intelligence Review* 39 (4): 261–283.

Kotsiantis, S. B., I. D. Zaharakis, and P. E. Pintelas. 2006. Machine learning: A review of classification and combining techniques. *Artificial Intelligence Review* 26 (3): 159–190.

Kotsiantis, S.B., I. Zaharakis, and P. Pintelas. 2007. Supervised machine learning: A review of classification techniques. *Emerging Artificial Intelligence Applications in Computer Engineering* 160 (1): 3–24.

Martinez, W. M. A 1995. Natural language processor with neural networks. In *1995 IEEE International Conference on Systems, Man and Cybernetics. Intelligent Systems for the 21st Century*, vol. 4, pp. 3156–3161. IEEE.

Mitchell, T. 1997. Introduction to machine learning. *Machine Learning* 7: 2–5.

Mitchell, T. 1996. *Machine Learning*. McCraw Hill, New York.

Mohri, M., A. Rostamizadeh, and A. Talwalkar. 2018. *Foundations of Machine Learning*. MIT Press, London, UK.

Montgomery, D. C., E. A. Peck, and G. G. Vining. 2012. *Introduction to Linear Regression Analysis*. Vol. 821. John Wiley & Sons, New Jersey, USA.

Muniyandi, A. P., R. Rajeswari, and R. Rajaram. 2012. Network anomaly detection by cascading k-Means clustering and C4. 5 decision tree algorithm. *Procedia Engineering* 30: 174–182.

Nasien, D., H. Haron, and S. S. Yuhaniz. 2010. Support Vector Machine (SVM) for English handwritten character recognition. In *2010 Second International Conference on Computer Engineering and Applications*, vol. 1, pp. 249–252. IEEE, Bali Island, Indonesia.

Odena, A. 2016. Semi-supervised learning with generative adversarial networks. *arXiv Preprint arXiv:1606.01583.*

Ojha, V. K., A. Abraham, and V. Snášel. 2017. Metaheuristic design of feedforward neural networks: A review of two decades of research. *Engineering Applications of Artificial Intelligence* 60: 97–116.

Panda, M., and M. R. Patra. 2007. Network intrusion detection using naive bayes. *International Journal of Computer Science and Network Security* 7(12): 258–263.

Pearson, K. 1901. LIII. On lines and planes of closest fit to systems of points in space. *The London, Edinburgh, and Dublin Philosophical Magazine and Journal of Science* 2 (11): 559–572.

Prasad, S. S., J. Kumar, D. K. Prabhakar, and S. Pal. 2015. Sentiment classification: An approach for Indian language tweets using decision tree. In *International Conference on Mining Intelligence and Knowledge Exploration*, pp. 656–663. Springer, Cham, Delhi, India.

Priddy, K. L., and P. E. Keller. 2005. *Artificial Neural Networks: An Introduction*. Vol. 68. SPIE Press, Washington DC, New York.

Ray, S. 2019. A quick review of machine learning algorithms. In *2019 International Conference on Machine Learning, Big Data, Cloud and Parallel Computing (COMITCon)*, pp. 35–39. IEEE, Delhi, India.

Rish, I. 2001. An empirical study of the naive Bayes classifier. In *IJCAI 2001 Workshop on Empirical Methods in Artificial Intelligence*, 3(22), pp. 41–46.

Roli, F., and G. L. Marcialis. 2006. Semi-supervised PCA-based face recognition using self-training. In *Joint IAPR International Workshops on Statistical Techniques in Pattern Recognition (SPR) and Structural and Syntactic Pattern Recognition (SSPR)*, pp. 560–568. Springer, Berlin, Heidelberg.

Rosenblatt, F. 1957. *The Perceptron, a Perceiving and Recognizing Automaton Project Para*. Cornell Aeronautical Laboratory, Ithaca, New York, USA.

Rumelhart, D. E., G. E. Hinton, and R. J. Williams. 1986. Learning representations by back-propagating errors. *Nature* 323 (6088): 533–536.

Russell, S., and P. Norvig. 2002. *Artificial Intelligence: A Modern Approach*, Prentice Hall, New Jersey.

Saimurugan, M., K. I. Ramachandran, V. Sugumaran, and N. R. Sakthivel. 2011. Multi component fault diagnosis of rotational mechanical system based on decision tree and support vector machine. *Expert Systems with Applications* 38 (4): 3819–3826.

Samuel, A. L. 1959. Some studies in machine learning using the game of checkers. *IBM Journal of Research and Development* 3 (3): 210–229.

Shahin, M. A., M. B. Jaksa, and H. R. Maier. 2001.Artificial neural network applications in geotechnical engineering. *Australian Geomechanics* 36 (1): 49–62.

Springenberg, J. T. 2015. Unsupervised and semi-supervised learning with categorical generative adversarial networks. *arXiv preprint arXiv:1511.06390.*

Sutton, R. S. 1992. Introduction: The challenge of reinforcement learning. In *Reinforcement Learning*, pp. 1–3. Springer, Boston, Brookline, MA.

Sutton, R. S., and A. G. Barto. 2018. *Reinforcement Learning: An Introduction.* MIT Press, London, UK.

Taghipour, N., and A. Kardan. 2008. A hybrid web recommender system based on q-learning. In *Proceedings of the 2008 ACM Symposium on Applied Computing*, pp. 1164–1168, New York, USA.

Theodoridis, S. 2008. *Pattern Recognition*. Elsevier B.V. p. 203, San Diego, California.

Turing, A. 1950. Computing Machinery and Intelligence, *Mind*, LIX (236): 433–460.

Watkins, C. J. C. H. 1989. Learning from delayed rewards, PhD Thesis, University of Cambridge, England.

Watkins, C. J. C. H., and P. Dayan. 1992. Q-learning. *Machine Learning* 8 (3–4): 279–292.

Willis, M. J., C. D. Massimo, G. A. Montague, M. T. Tham, and A. J. Morris. 1991. Artificial neural networks in process engineering. In *IEE Proceedings D (Control Theory and Applications)*, vol. 138, no. 3, pp. 256–266. IET Digital Library, New York, USA.

Wold, S., K. Esbensen, and P. Geladi. 1987 Principal component analysis. *Chemometrics and Intelligent Laboratory Systems* 2 (1–3): 37–52.

Wu, X., V. Kumar, J. R. Quinlan, J. Ghosh, Q. Yang, H. Motoda, G. J. McLachlan et al. 2008 Top 10 algorithms in data mining. *Knowledge and Information Systems* 14 (1): 1–37.

Yang, G. S., E. K. Chen, and C.W. An. 2004. Mobile robot navigation using neural Q-learning. In *Proceedings of 2004 International Conference on Machine Learning and Cybernetics (IEEE Cat. No. 04EX826)*, vol. 1, pp. 48–52. IEEE, New York, USA.

Yang, Y., and Q. Zhang. 1997. A hierarchical analysis for rock engineering using artificial neural networks. *Rock Mechanics and Rock Engineering* 30 (4): 207–222.

Yegnanarayana, B. 2009. *Artificial Neural Networks*. PHI Learning Pvt. Ltd., New Delhi, India.

Yu, Z., F. Haghighat, B. C. Fung, and H. Yoshino. 2010. A decision tree method for building energy demand modeling. *Energy and Buildings* 42 (10): 1637–1646.

Yu, W., X. Wang, F. Wang, E. Wang, and B. Chen. 2008. Notice of Retraction: The research of improved apriori algorithm for mining association rules. In *2008 11th IEEE International Conference on Communication Technology*, pp. 513–516. IEEE, New York, USA.

Zha, Z. J., Tao Mei, J. Wang, Z. Wang, and X. S. Hua. 2009 Graph-based semi-supervised learning with multiple labels. *Journal of Visual Communication and Image Representation* 20 (2): 97–103.

Zhan, Y., Z. Shi, and M. Liu. 2007. The application of support vector machines (SVM) to fault diagnosis of marine main engine cylinder cover. In *IECON 2007–33rd Annual Conference of the IEEE Industrial Electronics Society*, pp. 3018–3022. IEEE, New York, USA.

Zhang, H., A. C. Berg, M. Maire, and J. Malik. 2006. SVM-KNN: Discriminative nearest neighbor classification for visual category recognition. In *2006 IEEE Computer Society Conference on Computer Vision and Pattern Recognition (CVPR'06)*, vol. 2, pp. 2126–2136. IEEE, New York, USA.

Zhang, L., W. Dong, D. Zhang, and G. Shi. 2010. Two-stage image denoising by principal component analysis with local pixel grouping. *Pattern Recognition* 43 (4): 1531–1549.

Zhu, X. J. 2005. *Semi-Supervised Learning Literature Survey.* University of Wisconsin-Madison Department of Computer Sciences, Madison, Wisconsin, USA.

Zhu, X., and A. B. Goldberg. 2009. Introduction to semi-supervised learning. *Synthesis Lectures on Artificial Intelligence and Machine Learning* 3 (1): 1–130.

2 Detection and Prevention of Cancer through Artificial Intelligence and Machine Learning

Kaushik Dehingia
Gauhati University

Vishnu Narayan Mishra
Indira Gandhi National Tribal University

Bhagya Jyoti Nath
Barnagar College

CONTENTS

2.1 INTRODUCTION

Cancer originates from any part of human body through abnormal cell growth. The abnormal cell grows due to a genetic disorder. If we ignored this kind of disorder, it may become too dangerous for our health. At the earlier stage, surgery is the most suitable treatment for the cancer patient. Chemotherapy is widely used treatment in cancer diagnosis in which chemically synthesized drug is used. In radiotherapy, one type of energy called photon is used to destroy cancerous cells in the tumor site. Oncologists prominently use hormone therapy and immunotherapy in cancer treatment in present days as this treatment has lesser side effects than that others. Reducing the side effects of the treatment and duration of recovery is the prime objective for the medical researcher. Theoretical research has shown that a combination of two or more therapies helps to overcome this goal (Enderling and Chaplain 2014).

For a suitable and successful treatment, detection of cancer in an earlier stage is the main task. If the type and pattern of cancer are detected in the earlier stage, then the probability of curing the disease increases. If the disease is found in the earlier stages, then it helps minimise drug toxicity (Deshmukh 2020). Through the different screening processes, doctors detect cancer at the very earlier stages before cancer has grown up. For the early prediction of the outcome of cancer treatment, scientists develop new strategies day by day (Munir et al. 2019). Several computational methods like artificial intelligence (AI) and machine learning (ML) have revolutionized the field of healthcare and medicine with their wide-range applicability and prediction accuracy (Danee et al. 2017). An AI and ML technique helps to identify patterns and relationships of complex datasets (Xu et al. 2019). From observation of these patterns and relationship, one can be able to predict future outcomes of a cancer type effectively.

Mathematical modelling greatly helps to understand the dynamics of cellular interactions of a cancer patient (Kuznetsov et al. 1994). Appropriate models can predict survival and effective treatment for the patient. Machine learning algorithms are widely used to build a model by visualizing clinical data and estimating the parameter set of the model (Rockne and Scott 2019). This improves the prediction and accuracy of the model, which helps to forecasts the clinical outcome (Nagy et al. 2020, Shimizu and Nakayama 2020). In other words, ML algorithms aim to develop a mathematical model that fits the clinical data (Ho 2020).

2.2 TYPES OF CANCER

According to the American Cancer Society, there are 200 several types of cancer. Cancer can be classified based on its source, origin, place and size of the tumor. Smoking and tobacco, viruses and other infections and radiation are the main causes of cancer (Siegel et al. 2020).

1. The American Cancer Society reported in 2018 that 9320 mortalities occurred due to skin cancer among 3330 females and 5990 males in the USA (Siegel et al. 2020). For the case of melanomas patient, a country belonging from Asia, Africa and Latin America has a lesser ratio than that

of North America, Europe and Australia. Uncontrolled growth of abnormal cells in the skin leads to skin cancer. One kind of skin cancer is melanoma, which is the riskiest cancer. Nonmelanomas cancer begins on the skin with a dark spot. It can progress in size, irregular edges, itchiness, skin breakdown and color difference (Saba 2020). The mortality rate and risk factor of skin cancer will decrease if it is detected at the very beginning.

2. Breast cancer is a cancer that originates from female breast tissue. Majority of breast cancer occurs due to the lack of definite gene mutation (Tran et al. 2019). Infiltrating (invasive) ductal carcinoma, ductal carcinoma in situ, infiltrating (invasive) lobular carcinoma and lobular carcinoma in situ are four common breast cancers. At the very beginning of this cancer, it localizes to the milk ducts. Up to 2 cm of its size, cancer cannot spread up (Sahran et al. 2018). Therefore, early identification prevents from spreading on the body.
3. Abnormal cell growth in our brain tissue causes brain tumor. Primary brain tumor emerges in the brain and tends to stay there. Secondary brain tumors are those that first arise at some other parts of the body and then travel to the human brain (Saba 2020). Using MRI modality, brain tumors can be detected accurately.
4. Lung cancer is found in both males and females. Highest percentages of male and second highest for female death are because of lung cancer in the world according to data provided by the American Cancer Society in 2018 (Siegel et al. 2020). Lung cancer emerges from mutation of lung cells. Smoking causes about 90% of lung cancer cases.

Apart from these, other types of cancer like prostate cancer, liver cancer, blood cancer and eye cancer can be classified based upon their originating site. Using AI and ML, we can categorize different types of cancer and early detection through observing growth of cancer. This paper presents the application of some machine learning techniques in the field of cancer detection, diagnosis, prognosis and treatment.

2.3 TYPES OF CANCER GROWTH

A tumor develops from a single cell, and then it spreads up. Tumors that are not cancer are called benign. The growth rate of benign tumor is usually quite slow, and it has no spreading tendency. However, malignant tumors having high spreading tendency are made up of cancer cells and grow rapidly. The researcher takes many hypothesis and assumptions to describe the dynamics of cancer cells. They still have not clearly understood how a cancer cell birth takes place on the human body. Many papers (Forys and Marciniak-Czochra 2003, Talkington and Durrett 2015, Laird 1963, Sarapata 2013, Ogunrinde and Ayinde 2018, Friberg and Mattson 1997, Collins et al. 1956) worked on it and found that at the initial stage of the tumor, the malignant cell grows exponentially concerning time. However, once the malignant cell comes up interacting with the other healthy and immune cells of the body, its growth rates change time to time. Below we have discussed some mathematical laws, which have been taken by the researchers to describe cancer growth from its origin.

2.3.1 Exponential Growth

Exponential growth describes the ideal scenario in which cancer cells divide without constraint and continue to double indefinitely (Talkington and Durrett 2015). This law is appropriate for early tumor growth. For exponential growth, the size of tumor population P must satisfy the following ordinary differential equation:

$$\frac{dP}{dt} = CP \tag{2.1}$$

where t represent time and C be some constant (Figure 2.1).

2.3.2 Gompertz Growth

In 1825, Benjamin Gompertz first used this mathematical law to explain human mortality curves and the value of life insurances (Laird 1963). Anna Laird used it successfully to fit the growth of 19 tumor cell lines, and after that this law became popular in cancer research (Laird 1963). Usually, breast cancer follows this type of growth law. Gompertz law describes the tumor growth dynamics by the following differential equation:

$$\frac{dP}{dt} = r\log\left(\frac{K}{P}\right)P \tag{2.2}$$

where P is the size of the tumor population, r is the intrinsic growth rate and K is the maximum cell carrying capacity (Figure 2.2).

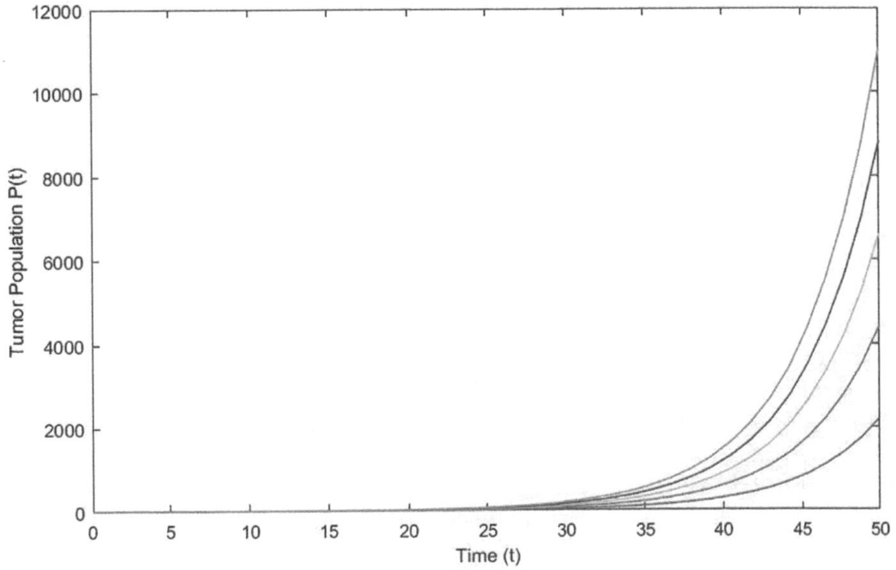

FIGURE 2.1 Exponential growth of tumor cell.

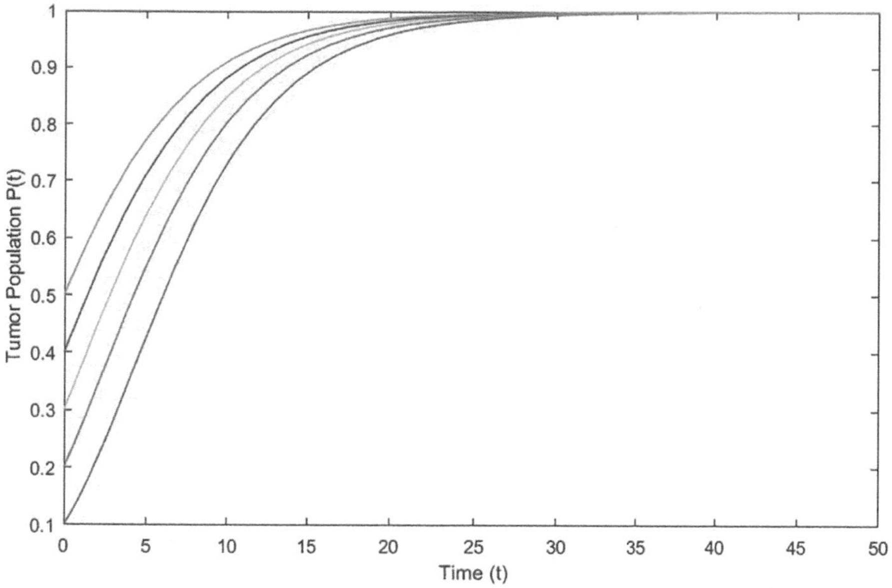

FIGURE 2.2 Gompertz growth of tumor cell.

2.3.3 Logistic Growth

Logistic growth is a simple law, which describes the complex cancer growth. In 1985, Krug and Taubert proposed this law to describe the growth of the Ehrlich ascites tumor (EAT) in a mouse (Forys and Marciniak-Czochra 2003). Logistical tumor growth means that the tumor will reach a maximum value, which is the carrying capacity, then either it will keep growing in the same volume or it will decay, but the tumor growth will not exceed the carrying capacity (AL-Azzawi et al. 2017). Kuznetsov et al. assumed in their research work (Kuznetsov et al. 1994) that the tumor grows logistically in the absence of treatment, and he used the logistic growth form $rP\left(1-\frac{P}{K}\right)$ to represent the growth of the tumor cells. The two parameters r and K represent the maximal growth rate and the inverse of tumor carrying capacity, respectively, and P is the size of the tumor population. The following differential equation can be used to describe tumor logistic growth (Figure 2.3).

$$\frac{dP}{dt} = rP\left(1-\frac{P}{K}\right) \tag{2.3}$$

Some researchers also used the tumor growth law such as power law, generalized logistic law, Bertalanffy law and stochastic Gompertz law to describe the growth of cancer within the human body.

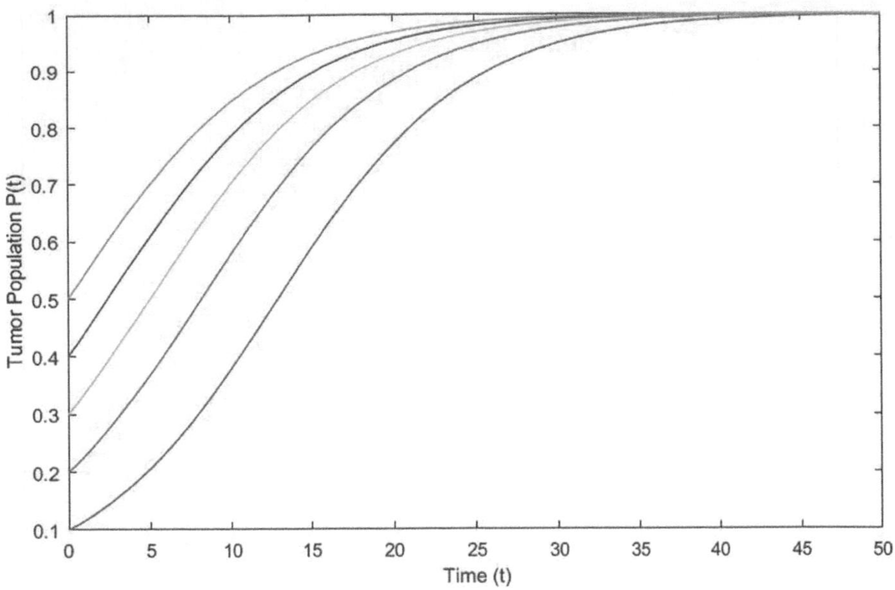

FIGURE 2.3 Logistic growth of tumor cell.

2.4 MACHINE LEARNING AND ITS TYPES

Artificial Intelligence is a computer science branch in which we discuss learning, planning and problem-solving performance of human intelligence. The main object of AI systems is to solve complex data-intensive problems. AI-based models are heavily used to predict outcome in advance and provide us with a deep insight into the effective treatment.

Machine learning (ML) is a branch of Artificial Intelligence where the general concept of inference is related to the trick of learning from data samples (Bishop 2006, Mitchell 2006, Witten and Frank 2005). Machine learning mainly refers to some algorithms, which are developed using the computer system, and they can map input data to some output predictions by learning from the training provided.

Two main types of Machine Learning techniques are Supervised and Unsupervised ML. The main difference between supervised and unsupervised learning is that to estimate the input data to the desired output, in supervised learning, a labeled set of training data is used, whereas no labeled training datasets are provided in the unsupervised learning.

Regression and clustering are two other types of ML process. In regression problems, the input data are mapped into a real-valued variable by a learning function. Thus, based on this process, the value of a predictive variable can be estimated for each new sample. Clustering is an unsupervised task where one tries to divide the data items into different categories or clusters to describe them. Thus, studying the characteristics that a new sample shares, it can be allocated to one of the identified clusters.

Another type of ML process is semi-supervised ML method in which both supervised and unsupervised MLs are combined. Mainly, in this process, both labeled and unlabeled datasets are combined for a better learning model.

Deep Learning (DL) in which deep neural networks are used to create a model is a subfield of machine learning. For image segmentation and detection, image phenotyping, clinical outcome predictions, etc. in oncology, deep learning techniques are widely used (Boldrini et al. 2019).

2.5 ML IN THE FIELD OF CANCER RESEARCH

With the current developments in the area of Artificial Intelligence (AI), detection and classification of different types of cancer in the human body using machine-learning techniques have become a new dimension of research. Different ML processes with supervised, unsupervised and deep learning techniques are used in detection and eradication process of different types of cancer like breast, brain, lung, liver, skin cancer, leukaemia, etc. in the human body. It helps in early detection and categorization of the patients into groups like low-risk or high-risk, etc. which is necessary for successful cancer research and treatment.

To develop different predictive models for effective and accurate decision-making in the field of cancer research, various techniques like (i) Artificial Neural Networks (ANNs), (ii) Bayesian Networks (BNs), (iii) Support Vector Machines (SVMs) and (iv) Decision Trees (DTs) have been widely applied by the researchers. Among these, most recent ML approaches are SVMs, which are applied in the field of cancer research for its prediction.

In the past decades, different research works have been published where ML techniques are used to predict cancer susceptibility, recurrence and survival. ML techniques like ANNs and DTs have been used in cancer detection for the last three decades (Bottaci et al. 1997, Maclin et al. 1991, Simes 1985). But in the last decade, researchers have started to use supervised ML techniques like SVMs and BNs in cancer prediction and diagnosis (Akay 2009, Chang et al. 2013, Chuang et al. 2011, Eshlaghy et al. 2013, Exarchos et al. 2012, Kim and Shin 2013). The role of different ML techniques in various stages of cancer treatment is discussed below.

2.5.1 Role in Screening and Diagnosis

Detection of cancer at the early stage and then initiation of its treatment in a proper manner is a significant task for oncologists to reduce morbidity and mortality of cancer patients. There are different screening processes that fail in detecting treatable malignancies, or some are unnecessary. Therefore, to mitigate the harms of existing methods and increasing the benefits of existing methods, we may improve screening and diagnosis using some ML applications. For example, females of the age range 50–70 are advised for scanning in every three years to detect breast cancer at the possible earliest stage (Cancer Research UK 'Breast Screening'. webpage 2017). From such larger number of scanning, AI techniques can be used to find out the suspicious images. This process will help to reduce the time of detection of the suspicious images, and it will reduce unnecessary diagnoses for the others. To detect suspicious scanned images, which may have tumors, an algorithm of groups of different AI algorithms and discrete wavelet transform (DWT) was presented by Lei Zheng and Chan (2001). Also after its early-stage detection, the development of breast cancer can be predicted using ML techniques. For this purpose, Mc Kinney et al. (2020) proposed a model, which predicts the development of breast cancer for 2 years.

In many screening and diagnosis processes of cancer research, radiographic and pathologic data are involved in different medical images. Among these, Convolutional Neural Network (CNN) is mainly used in image analysis. Huang et al. (2019) developed a superior CNN-based method for lung cancer screening in the year 2019. They used a group of 25,097 testing data and 2350 validating data to propose this model. CNN-based algorithms are also used for lung cancer detection (Xie et al. 2019, Shen et al. 2017). For this, CNN algorithm divides two parts: (i) feature extractor, which extracts the most relevant features from a single patch and (ii) classifier, in which the aim is to classify the extracted features. Similarly, different ML techniques are used for detection of other types of cancer such as skin cancer, leukaemia, liver cancer, etc. (Chaudhary et al. 2018).

2.5.2 Role in Prognostication and Risk Stratification

Cancer is a disease, which has a chance of recurrence. Therefore, researchers always try to find out a strategy to find the probability of future risk. Traditional techniques used different post-operative histopathology characters like size, grade, number of lymph nodes involved, local infiltration status, etc., which indicate reappearance of cancer in post-operative cases. But AI techniques cover the other features of a patient like sex, age, geographical inconsistencies, the historical backgrounds and performance status of the disease, etc. By utilizing all these additional data, AI-algorithms can be used for more precise estimation of the future risk. Already, different works (Chaudhary et al. 2018, Chen and Millar 2017) have been done in these regards. Nowadays, chemists and pharmacists also use ML for representations of structure of a chemical drug. They invented personalized medicine for a specific patient by modeling under consideration of cellular interaction, microbial and tumor drug sensitivity and patient condition.

2.5.3 Role of Deep Learning in Image Construction and Radiation Oncology

Deep learning is mainly used in radiation oncology because it has a power of analyzing unstructured data, and it can extract the suitable features without any human guidance. CNN, auto-encoder, deep belief network, deep de-convolution neural network, feed forward neural network and transfer learning are used in this regard (Rattan et al. 2019). Among these, some neural networks such as deep belief networks and fully connected feed-forward neural networks were used in early studies of radiation oncology. But, with the developments of deep learning, CNN, auto-encoder and transfer learning are applied to radiation oncology since 2017 (Figure 2.4).

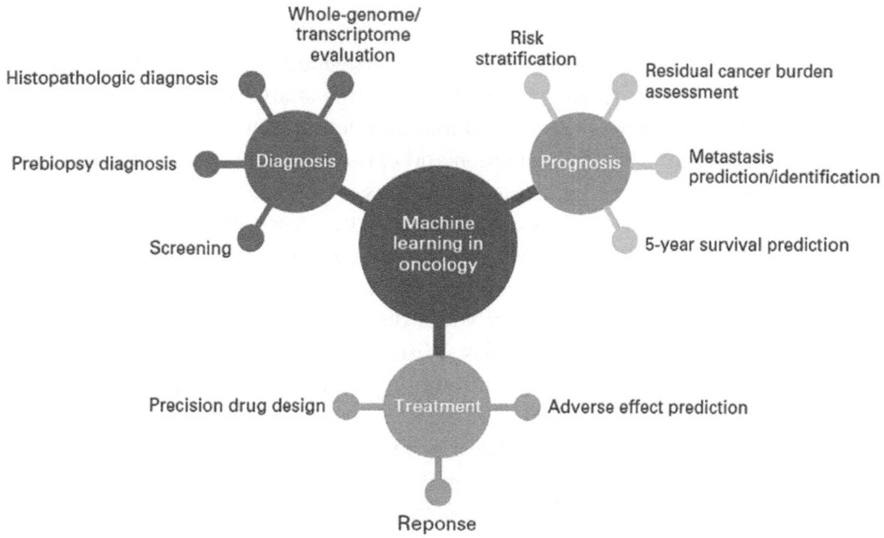

FIGURE 2.4 ML techniques used at different stages in oncology research (Nagy et al. 2020).

Though different deep learning techniques are used in radiation oncology, all of them have a similar framework. Suppose, X be a training dataset like histology images, CT scans, etc., and Y be their target labels, a deep learning method maps a discriminant function $f : X \rightarrow Y$ such that there exists a Y for a given X. Since a neural network with a single layer of infinite width can approximate any continuous functions (Kurt et al. 1989), a neural network is usually used as a discriminant function. Further, f is determined by the parameters used in the model, defining a loss function and applying an optimization process.

2.5.4 Role in Treatment

Since there are large data of different cancer patients and their responses to different treatment strategies of cancer, this information can be used with the help of proper ML techniques to predict the proper treatment for a new patient. This will speed up the treatment in a proper way, which is vital to control cancer growth. In radiation oncology and interventional radiology, different ML approaches are used to find proper dosing strategies, and it also helps in assessing responses to the therapy (Rattan et al. 2019). These approaches are based on CNNs, radiomics or their combination, which also account for patient-specific anatomy, and hence, accurate prediction of adverse effects may be possible for these ML approaches. As a result, treatment policy can be planned in the proper time. Another feature of ML technique is analyzing gene expression data of a patient which is used to develop a supportive clinical decision for prediction of accurate survival time, and so patients get effective treatment according to their survival. This could prevent unnecessary surgical treatment time and the support patients' normal health.

2.6 CONCLUSION

Over the last few decades, cancer research has grown up because it is the most common cause of deaths in developed countries. In this study, we have focused on the application of artificial intelligence and machine learning to cancer for saving lives in the upcoming era. Advanced detection and classification of cancer may save many lives, and this can be done by using the various techniques of AI & ML. We have discussed various types of cancer growth pattern because it plays a vital role in cancer detection and classification. ML algorithms automatically build an analytical model for given input data, which can help improve patient health and personalized medicine. Techniques like CNN, ANN and deep learning play a significant role in radiation oncology and cancer genomics. With the help of these techniques, radiation oncologists can develop a suitable radiotherapy treatment by improving treatment quality, speed and patient survival.

Different AI-based models have been proposed, but few of them are performing well in cancer detection and classification. Designing a good model is the main task for the scientists and researchers in this field. For this, the researcher should have to understand the cancer growth phenomena, the dynamics of cellular interaction, treatment procedure and the innovative method of data analysis.

REFERENCES

Akay, M.F. 2009. Support vector machines combined with feature selection for breast cancer diagnosis. *Expert Syst Appl.* 36: 3240–7.

AL-Azzawi, S.N., F.A. Shihab and M.M. Al-Sayyid. 2017. Solution of modified Kuznetsov model with mixed therapy. *Global J Pure Appl Math.* 13(9): 6269–6288.

Bishop, C.M. 2006. *Pattern Recognition and Machine Learning.* New York: Springer.

Boldrini, L., J.E. Bibault, C. Masciocchi, Y. Shenand and M.I. Bittner. 2019. Deep learning: A review for the radiation oncologist. *Front Oncol.* 9: 977. Doi: 10.3389/fonc.2019.00977.

Bottaci, L., P.J. Drew, J.E. Hartley, et al. 1997. Artificial neural networks applied to outcome prediction for colorectal cancer patients in separate institutions. *Lancet.* 350: 469–72.

Cancer Research UK 'Breast Screening'. Webpage 2017.

Chang, S.W., S. Abdul-Kareem, A.F. Merican and R.B. Zain. 2013. Oral cancer prognosis based on clinicopathologic and genomic markers using a hybrid of feature selection and machine learning methods. *BMC Bioinforma.* 14: 170.

Chaudhary, K., O.B. Poirion, L. Lu, and L.X. Garmire. 2018. Deep learning-based multi-omics integration robustly predicts survival in liver cancer. *Clin Cancer Res.* 24: 1248–59. Doi: 10.1158/1078-0432.CCR-17-0853.

Chen, Y., and J.A. Millar. 2017. Machine learning techniques in cancer prognostic modelling and performance assessment. *Frontiers of Biostatistical Methods and Applications in Clinical Oncology*, Springer, Singapore, 193–230.

Chuang, L.Y., K.C. Wu, H.W. Chang and C.H. Yang. 2011. Support vector machine-based prediction for oral cancer using four SNPs in DNA repair genes. *Lecture Notes Eng Comput Sci.* 2188(1): 426–429.

Collins, V.P., R.K. Loeffler and H. Tivey. 1956. Observations on growth rates of human tumors. *Am J Roentgeol Radium Ther Nucl Med.* 76: 988–1000.

Danee, 2P., R. Ghaeiniand and D.A. Hendrix. 2017. A deep learning approach for cancer detection and relevant gene identification. *Pacific Symposium on Biocomputing*, Kohala Coast, Hawaii, USA.

Deshmukh, S.K. 2020. Artificial intelligence and machine learning in cancer care: Current applications and future perspectives. *J Cancer Immunol.* 2(2): 35–36.

Enderling, H. and M.A.J. Chaplain. 2014. Mathematical modeling of tumor growth and treatment. *Curr Pharm Des.* 20(30): 4934–4940. Doi: 10.2174/1381612819666131125150434.

Eshlaghy, A.T., A. Poorebrahimi, M. Ebrahimi, A.R. Razavi and L.G. Ahmad. 2013. Using three machine learning techniques for predicting breast cancer recurrence. *J Health Med Inform.* 4: 124.

Exarchos, K.P., Y. Goletsis and D.I. Fotiadis. 2012.A multiscale and multiparametric approach for modeling the progression of oral cancer. *BMC Med Inform Dec Mak.* 12: 136.

Forys, U. and A. Marciniak-Czochra. 2003. Logistic equations in tumour growth modelling. *Int. J Appl Math Comput Sci.* 13(3): 317–325.

Friberg, S. and S. Mattson. 1997. On the growth rates of human malignant tumors: Implications for medical decision making. *J Surg Oncol.* 65: 284–297.

Ho, D. 2020. Artificial intelligence in cancer therapy. *Science,* 367 (6481): 982–983. Doi: 10.1126/science.aaz3023.

Huang, P., C.T. Lin, Y. Li et al. 2019. Prediction of lung cancer risk at follow-up screening with low-dose CT: A training and validation study of a deep learning method, *Lancet Digit Health,* 1: E353–E362.

Kim, J. and H. Shin. 2013. Breast cancer survivability prediction using labeled, unlabeled, and pseudo-labeled patient data. *J Am Med Inform Assoc.* 20: 613–8.

Kurt, H., S. Maxwell, W. Halbert et al. 1989. Multilayer feed forward networks are universal approximators. *Neural Net.* 2(5): 359–366.

Kuznetsov, V.A., I.A. Makalkin, M.A. Taylor and A.S. Perelson. 1994. Nonlinear dynamics of immunogenic tumors: Parameter estimation and global bifurcation analysis. *Bullet. Math. Biol.* 56(2), 295–321.

Laird, A.K. 1963. Dynamics of tumor growth. *British J Cancer.* 18:490–502.

Maclin, P.S., J. Dempsey, J. Brooks and J. Rand.1991. Using neural networks to diagnose cancer. *J Med Syst.* 15: 11–9.

Mc Kinney, S.M., M. Sieniek, V. Godbole et al. 2020. International evaluation of an AI system for breast cancer screening. *Nature.* 577: 89–94.

Mitchell, T.M. 2006. *The Discipline of Machine Learning: Carnegie Mellon University.* Carnegie Mellon University, School of Computer Science, Machine Learning Department, Pittsburgh, PA.

Munir, K., H. Elahi, A. Ayub, F. Frezza and A. Rizzi. 2019.Cancer diagnosis using deep learning: A bibliographic review. *Cancers.* 11, 1235; Doi: 10.3390/cancers11091235.

Nagy, M., N. Radakovich and A. Nazha. 2020. Machine learning in oncology: What should clinicians know? *JCO Clin Cancer Info.* Doi: 10.1200/CCI.20.00049.

Ogunrinde, R.B., and S.O. Ayinde. 2018. Interpolating and Gompertz function approach in tumour growth analysis. *Am J Math Statistics.* 8(5): 119–125.

Rattan, R., T. Kataria, S. Banerjee, et al. 2019. Artificial intelligence in oncology, its scope and future prospects with specific reference to radiation oncology. *BJR Open* 1: 20180031.

Rockne, C.R. and J.G. Scott. 2019. Introduction to mathematical oncology. *JCO Clin Cancer Info.* Doi 10.1200/CCI.19.00010.

Saba, T. 2020. Recent advancement in cancer detection using machine learning: Systematic survey of decades, comparisons and challenges. *J Infect Pub Health.* 13: 1274–1289.

Sahran, S., A. Qasem, K. Omar, et al. 2018. Machine learning methods for breast cancer diagnostic. *Breast Cancer Surg.* Doi: 10.5772/intechopen.79446.

Sarapata, E. A. 2013. A comparison and catalog of intrinsic tumor growth models. *HMC Senior Theses.* 52. https://scholarship.claremont.edu/hmc_theses/52.

Shen, W., M. Zhou, F. Yang, et al. 2017.Multi-crop convolutional neural networks for lung nodule malignancy suspiciousness classification. *Pattern Recog.* 61: 663–73.

Shimizu, H. and K. Nakayama. 2020. Artificial intelligence in oncology. *Cancer Sci.* 111: 1452–1460 Doi: 10.1111/cas.14377.

Siegel, R.L., K.D. Miller and A. Jemal. 2020. Cancer statistics. *CA Cancer J Clin.* 70: 7–30.

Simes, R.J. 1985. Treatment selection for cancer patients: Application of statistical decision theory to the treatment of advanced ovarian cancer. *J Chronic Dis.* 38: 171–86.

Talkington, A. and R. Durrett. 2015. Estimating tumor growth rates in vivo. *Bull Math Biol.* 77(10): 1934–1954.

Tran, W.T., K. Jerzak, F.I. Lu, et al. 2019.Personalized breast cancer treatments using artificial intelligence in radiomics and pathomics. *J Med Imag Radiation Sci.* 50: S32–S41.

Witten, I.H., and E. Frank. 2005. *Data Mining: Practical Machine Learning Tools and Techniques.* Morgan Kaufmann, UK.

Xie, H., D. Yang, N. Sun, Z. Chen and Y. Zhang. 2019. Automated pulmonary nodule detection in CT images using deep convolutional neural networks, *Pattern Recog.* 85: 109–19.

Xu, Y., A. Hosny, R. Zeleznik, et al. 2019. Deep learning predicts lung cancer treatment response from serial medical imaging. *Clin Cancer Res.* 25: 3266–75, Doi: 10.1158/1078-0432.CCR-18-2495.

Zheng. L. and A.K. Chan. 2001. An artificial intelligent algorithm for tumor detection in screening mammogram. *IEEE Trans Med Imaging.* 20: 559–67. Doi: 10.1109/42.932741.

3 A Review on Soft Computing Techniques in Nanomagnetism and Its Impact on Biomedical Applications

Korobi Konwar and Pritam Deb
Tezpur University

CONTENTS

3.1 INTRODUCTION

The extensive utilization of soft computing methods in several application fields makes it an essential tool for the advancement of digital technology implied for the benefit of mankind. From the last two decades, scientific community has started to devote efforts to corroborate the Soft Computing techniques with material science (Ibrahim 2016, Chandrasekaran et al. 2010). However, traditionally used hard computing control systems have limitations to obtain accurate solutions for existing complex problems by using few numerical modelling, symbolic logic reasoning, etc. (Zarei et al. 2009, Yue and Jia 2013). On the contrary, the soft computing technique involves approximate modelling with advanced landscapes of nonlinear programming, artificial intelligence as a lenient of uncertainty and inaccuracy. The complexity residing in the modern machinery is an unavoidable problem, which can be solved by considering the soft computing techniques (Attarzadeh and Ow 2010). Soft computing is a combination of various methods which include fuzzy logic, algorithm, probabilistic approximation, artificial neural networking, etc. (Nikravesh 2008, Abioduna et al. 2018, Pezzella et al. 2008). These methods can be implemented together to solve a complex real-life problem by manipulating the uncertain issues and imprecision.

Recently, soft computing is attaining a remarkable position due to its applicability toward nanomagnetic device fabrication along with its efficient pathway for the development of smart biomedical techniques (Suneetha et al. 2020, Vashist et al. 2012). Indeed, nanotechnology has a generous influence on the manufacturing industry. Hence, it is essential to recognize and estimate the critical factors required for the applicability of nanotechnology. It is observed from the reported works (Shariati et al. 2017, Azimi et al. 2011, Abioduna et al. 2018) that analytical network process (ANP) method is the powerful element for the betterment of nanotechnology-based applications. ANP is a subdivision of the multicriteria decision making technique. During the formation procedure of a decision-making problem, ANP model considers tangible as well as intangible criteria so that it can control all kinds of relationship irrespective of its dependency. Nevertheless, intuitionistic fuzzy set (IFS) (Konwar et al. 2019) is also a renowned method where intrinsic ambiguity residing in a model processing is considered during designing of a decision-making problem. There are few reported works where IFS and ANP are utilized together to estimate the critical factors for the applicability of nanotechnology in the manufacturing industries. The IFS-ANP technique has significant potential for attaining the criteria weight in an ambiguous environment during the procedure of problem modelling. Although soft computing is addressed in bio-image processing (Agrawal et al. 2014, Kaur et al. 2020, Altameem et al. 2014), it is an effectual methodology to control the uncertainties found in image data. In addition, incorporation of swarm intelligence, cellular automata and also the genetic algorithms can facilitate the systems with required promising properties such as advanced complex networking, growth, self-repair, etc. (Reddy et al. 2020). The aforementioned soft computing techniques are executed successfully by the scientific community in real-time complications such as control systems in industrial plants, air traffic device, etc. Moreover, from last two decades, various approaches (Agrawal et al. 2014, Kaur et al. 2020, Altameem et al. 2014) are employed to attain better biomedical image such as implementation of fuzzy clustering to obtain Magnetic Resonance (MR) image segmentation of human brain, execution of fuzzy connectedness to the segmentation of biomedical images, employment of fuzzy methods and statistical plans for system recognition and allocation, etc. There are several advanced approaches that are found recently for image processing. A method is found where artificial fish swarm algorithm is combined together with fuzzy c-means to demonstrate the MR image quality of human brain (Ma et al. 2015). On the other hand, another work has been established by introducing a counter extraction model in order to perform segmentation of deep cervical flexor muscles during ultrasound imaging process (Kim et al. 2015). A dual-threshold method is also found for segmentation of white blood cells (Yan et al. 2015). Convolutional neural network technique is observed to establish in order to segmenting of brain tumor (Bhandari et al. 2020). As nanomagnetism is used to obtain advanced diagnostic tools, nanomagnetic and soft computing techniques together will enhance the device efficiency.

In this chapter, we have evaluated various techniques involved in soft computing and their execution on material science and nanotechnology. The advantages of soft computing tools' implementation in nanomagnetism device fabrication are reviewed critically, and the residing limitations are discussed thoroughly. Nanomagnetic devices with soft computing tools also develop a favorable pathway to find out a solution for the existing lacuna of medical science by addressing advanced diagnostic tools.

3.2 TYPES OF SOFT COMPUTING TECHNIQUES

In this section, we have discussed various kinds of soft computing techniques. Soft computing (SC) covers several methods, which contain artificial neuro-computing, chaotic methods, fuzzy logic, probabilistic approximation, Bayesian network, genetic algorithms, evolutionary computation technique and computational learning theory. Unlike analytical approaches, (which include deterministic reasoning model, binary logic and crisp logic), SC deals with computing models that are able to solve the vague and uncertain problems. SC techniques can take over the nonlinearity, which results in computational simplicity over analytical techniques. Such methods can handle a huge number of information smoothly and can be considered to mimic biological systems during machine learning (Ibrahim 2016, Chandrasekaran et al. 2010, Zarei et al. 2009, Yue and Jia 2013, Attarzadeh and Ow 2010).

Among all the techniques, the underlying motivation on the development of Artificial Neural Network (ANN) is to replicate the neurological activities of biological systems (Abioduna et al. 2018, Bhandari et al. 2020). ANN is arranged for definite uses, such as array recognition, climate forecast, image processing, intelligent security, etc. The aim of ANN is to make the traditional computing systems a bit closer to the human brain activities. The nonlinear correlation between input and output results best outcome in ANN. ANN is an appropriate tool to overcome the limitations when there is absence of algorithm or particular set of guidelines to be followed, so as to find a solution to the problem. Moreover, the connectivity of nonlinear computational segments, known as neurons, is found in ANN (Kim et al. 2015). Each neuron implements minute operations, and the net operational contribution of entire neurons is considered as weighted combination of entire operations. A neural network is trained to produce the required outputs by providing teaching patterns and allowing the network to alter the weight function conforming to formerly define supervised learning or unsupervised learning guidelines. If the outcome pattern can be known for the precise input, then the network is said to be trained by supervised learning, whereas, in case of unsupervised learning, the output of the considered network is found to be trained to replicate the patterns of the provided input (Bhandari et al. 2020). However, the results obtained from ANN are mostly dependent on the accurateness of the provided data. Moreover, ANN has successful applicability in the field of predictions.

The idea about fuzzy logic was presented by Zadeh (Nikravesh 2008). Fuggy logic is a process of demonstrating the inherent knowledge of human beings. Generally, it is a simplified representation of binary logic. In contrast to binary logic, a range of (0; 1) for truth value is found to consider the variables in case of fuzzy logic. There are various areas where fuzzy logic is executed and few of those are: pattern recognition, biomedical applications, climate forecasting, industrial control systems, etc. As fuzzy logic affords a good computational model in order to encompass the ability of binary logic, it can be considered as a powerful candidate to execute the complex limitations of material science. Nevertheless, fuzzy logic bids a practical approach for scheming nonlinear control systems by overcoming the limitation of traditional techniques.

However, Bayesian network (BN) provides Bayesian inference during probabilistic computation (Agrahari et al. 2018). It is a traditionally used probabilistic method to execute in machine learning techniques, which can solve the difficulties associated with system modeling and uncertain decisions of several domains. BN is a probabilistic graphical language that can be considered as a suitable candidate for knowledge illustration and reasoning. BN plays an important role in preprocessing, learning and postprocessing of data in machine learning. BN can deal with inadequate datasets and draw a connection between past knowledge and patterns acquired from data. Moreover, another technique used in soft computing is computational learning theory, which studies designing of a computer programming. It has the capacity to identify the limitation of learning by machines.

In order to solve several real-time optimization problems, genetic algorithm (GA) models (Pezzella et al. 2008) are executed which belong to the artificial intelligence. The fundamental concept of GA is to copy the natural selection to attain an appropriate selection for the desired application. GA is generally a machine learning model motivated by natural progression and used in order to find a solution for complicated search problems of engineering. GA is found to have efficient applicability in the field of biomedical engineering, games theory, control systems engineering, designing of electronics, etc. These techniques are employed in various aspects during the fabrication of nanomagnetic devices which is discussed in the next section.

Indeed, the evolutionary computation technique has considered to be a significantly beneficial approach in order to solve optimization problems with its capability to traverse through complicated problem spaces. Furthermore, this technique can be utilized as a creative element to produce new designs. The presence of such duality results in a better applicant to be integrated into a procedure for scheming and physically constructing self-assembling systems. The concept of the design space is very vital in the case of evolutionary computational model. Consideration of a vague space can affect the software performance in addition to preventing the formation of self-assembled system.

Such soft computing techniques are observed as useful tools to execute in various fields in order to enhance the applicability of the fabricated devices. As scientific community is concerned about better device fabrication in technology aspects, various soft computing techniques are implemented in the material science research. Few of such kind of works are explained in the next section.

3.3 SOFT COMPUTING IN MATERIAL SCIENCE AND NANOMAGNETISM

Nanotechnology has certainly assured great developments for human requirements. The incorporation of the aforementioned soft computing techniques on nanomagnetism brings a new way to develop smart devices with efficient applicability. The utilization of image processing technique and computer graphics in nanomanipulators is providing an interactive system interface to the microscopic instruments, which provides an opportunity to examine and manipulate the atomic-scaled surface. This will also help to understand the structural property of nanoscaled magnetic smart devices. Moreover, in molecular nanotechnology, the genetic algorithm technique has been executed in order to design an automatic system (Falvo et al. 1997, Shrivastava and Dash 2009). However, swarm intelligence algorithm is also enthused by combined intelligence of social living organisms. The collaborative behaviors obtained from the distinct interactions obey the self-organization process (Whitesides and Grzybowski 2002, Cai et al. 2007, O'Grady et al. 2008, O'Grady et al. 2010, Lee et al. 2013). But such collective nature cannot occur always from direct individual interaction. As an alternative, indirect stigmergy is observed to be employed. Such swarm intelligence is used to aid human behavior investigation by critically noticing other social creatures in order to solve several optimization difficulties. However, swarm intelligence is recognized as a substantial technique for parallel processing and instantaneous controlling of the inputs to yield an expected emergent result (Bonsma et al. 2003). In case of nanoscaled magnetic device fabrication, where no mathematical modelling exists, such intelligent techniques can be applied to regulate intelligently the development of the device. As the physical behavior of nanoscaled magnetic materials is found as promising for efficient applicability, researchers are trying to solve the dynamic behaviors of such materials using soft computing methods.

Some of the specific applications of SC techniques used in material science are explained in the tabular representation as shown in Table 3.1.

Such applications found in soft computing techniques need more research to come up with significant outcomes. However, most of the techniques that are found useful for biomedical applications are explained in detail in the next section.

3.4 IMPACT OF SOFT COMPUTING ON BIOMEDICAL APPLICATIONS

Contemporary digital technologies have delivered a platform to the scientific community to develop smart tools in order to support the medical decision making. Nanotechnology provides assurance to achieve emerging benefits for biomedical research with improved quality for diagnostic tools, growth of human immunity system by medical nanomachines, rejuvenation of tissues, age tackling tools, etc. (Agrawal et al. 2014, Kaur et al. 2020, Altameem et al. 2014). It is claimed by the researchers that the implementation of nanotechnology toward biomedical field provides ultimate benefits to the human life by overcoming the limitations of medical tools as shown in Figure 3.1. The nanomaterials and nano-scaled devices are used for designing advanced biosensors, immunoisolation treatments and advanced drugs (Kaur et al. 2020, Altameem et al. 2014). However, nanoscaled magnetic materials are also used to develop various smart devices and also tested for several potential areas such as tagging magnetic nanoparticles with the help of quantum dot as a form of biological markers and also as a smart drug which starts to activate only for certain required circumstances (Reddy et al. 2020, Ma et al. 2015, Kim et al. 2015). Furthermore, scientific community has found a technique to regulate the size of compactly packed DNA patterns, among nature's effectual paths for transferring

TABLE 3.1
Tabular Representation of SC Techniques with Applications and Specifications

Soft Computing Techniques	Type of Applications	Remarks	References
• Artificial neural network	• Gas recognition system • Self-assembling robots	• ANN was implemented in order to model signal sensitivity of an array, and a gas recognition system was employed to classify as well as identify explosive gases as well as detection of volatile organic compounds. Principal component analysis was used to analyze the characteristics of multi-dimensional signals received from the nine developed sensors. An error back propagation (EBP) learning algorithm was employed in the multi-layer neural network which was used for implementation of gas pattern recognizer. • A three-layered BP-ANN model was implemented in order to detect the purity of SrTiO3 nanocrystals. The used network was modified by implementing reaction period, temperature and respective molar ratio of the precursors. It was observed that BP neural network is effective for forecasting the purity of $SrTiO_3$ nanosystems. • An Evolvable Neural Interface (ENI) is used by developing an outline to recognize the autonomous intelligent system. Consequently, the interface provides collaboration between two systems such as higher-level and lower-level neural system, respectively. Higher-level neural system is considered for elementary purpose actions, whereas lower-level neural system is considered for solving real-life complications. It is expected that each autonomous element will be adept of adjusting itself for the mission, and the structures of ANTS will be built on carbon nanotube elements. • Self-assembling robot development can attain a notable applicability for fabrication of advanced nanoscaled material. Self-assembling is required essentially for advanced nanorobot development. Self-assembling elements can be brought easily to the hazardous location than of a complete robot. A self-assembled device can get repaired by itself if it gets damaged by regenerating the destroyed parts.	• Lee et al. (2000) • Lee et al. (2002) • Qing-li and Quan-xi (2006) • O'Grady et al. (2008)

(*Continued*)

TABLE 3.1 (*Continued*)
Tabular Representation of SC Techniques with Applications and Specifications

Soft Computing Techniques	Type of Applications	Remarks	References
• Swarm intelligence	• Bio-nano robots • Nanobots • DNA computers	• Bio-nano robots are developed from biological elements such as DNA structures or proteins in the nanoscaled size with high efficacy for therapeutic treatments. In order to execute the desired tasks to achieve the goal such as detection of cancerous cell, nanorobots can connect and organize as a group in a collective manner following the swarm intelligence technique. • A DNA nanomechanical device was developed which can mimic the ribosome transitional capabilities in order to facilitate positional synthesis DNA molecules in a definite DNA sequence. It is observed that the developed device can be executed as a variable input system for DNA computation and can be considered in information encryption. • By considering the idea of artificial intelligence, Autonomous Nanotechnology Swarm (ANTS) architecture allots the autonomous elements into swarms and arranges them in the form of hierarchy. • Another novel swarm algorithm named 'Perceptive Particle Swarm Optimization' (PPSO) is addressed which is an augmentation of the traditional Particle Swarm Optimisation (PSO) algorithm for physical-world applications. PPSO algorithm is designed unambiguously with nanotechnology for the first time in scientific community. Each and every element of PPSO algorithm is analyzed properly, and the design of the algorithm is considered in such a way so that a larger number of elements can be considered. As a consequence, this algorithm is appropriate for monitoring various agents of nanotechnology agents to execute their tasks efficiently.	• Chandrasekaran and Hougen (2006) • Liao and Seeman (2004) • Whitesides and Grzybowski (2002) • Cai et al. (2007)

(*Continued*)

TABLE 3.1 (*Continued*)
Tabular Representation of SC Techniques with Applications and Specifications

Soft Computing Techniques	Type of Applications	Remarks	References
• Fuzzy Logic (FL)	• DNA nanotechnology	• Though FL is implied in limited material engineering works, it is observed that most of the works follow the same procedures that are used for not only in membership function generation, but also for defuzzification module designing. In order to design the material properties and selection, fuzzy logic is applied to make the development procedure more expressive along with an easily interpretable solution. • Fuzzy logic has an advantage over ANN as it provides more information than that of ANN. Moreover, in few of the published works, it is found that, if the total number of input variables is large enough, then it results in validation for fuzzy logic application to obtain an easily interpretable solution. • The pattern of defuzzification procedure relies on experience and accomplishes using a trial-and-error technique. Defuzzification procedure can provide counter-intuitive outcomes even when it has been designed very critically • Fuzzy reasoning technique is found to realize by considering DNA computing to add a novel path to the fuzzy reasoning technique in the nanoscale computing. The synthetic DNA sequence was fuzzified by considering quantum dot and the fundamental method of fuzzy reasoning was understood on a DNA chip by considering fuzzy DNA. Such DNA computing based fuzzy reasoning can be executed in various fields of pattern classification, weather prediction, object recognition, etc. • Fuzzy logic-based model can deliver a paradigm to accomplish some type of fuzziness found in human reasoning. Therefore, the applicability of fuzzy logic to material science should not be misinterpreted as conclusive in conception.	• Lee et al. (2013) • Zadegan et al. (2015) • Shariati et al. (2017) • Fahim et al. (2019)

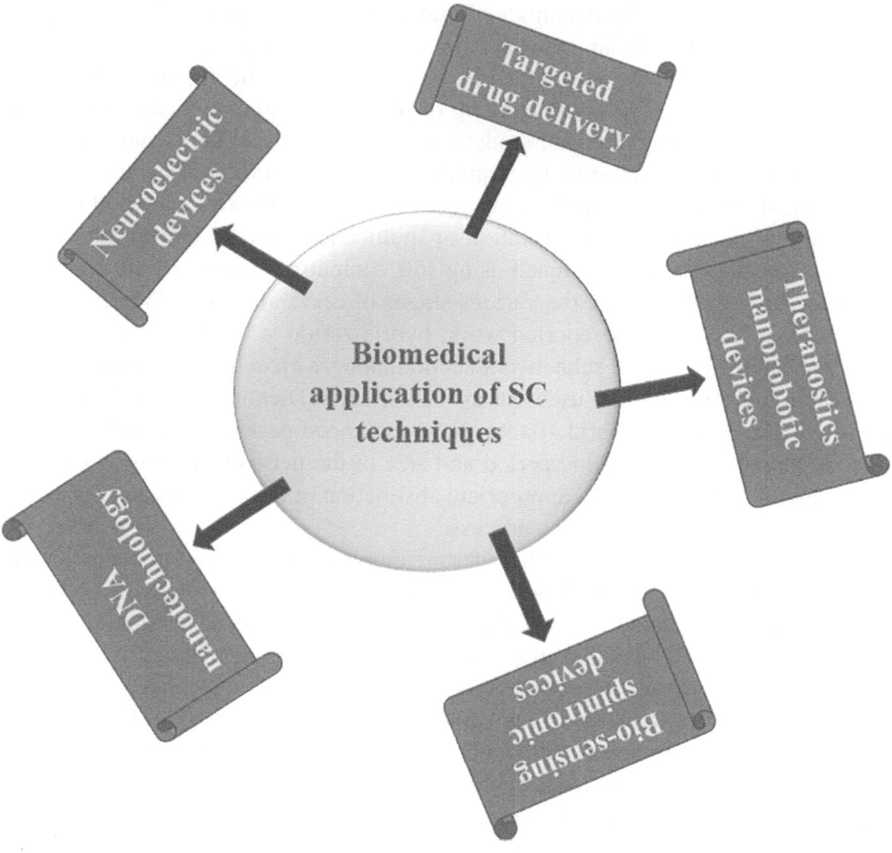

FIGURE 3.1 Schematic representation of applications of SC techniques in biomedical field.

the information of human gene. Such method helps to develop the effectiveness of gene therapy and disease preclusion (Yan et al. 2015, Bhandari et al. 2020). When nanocomputers and nanorobots will be incorporated to the medical facilities, the nanomedicines will reach the next level of technology with enhanced effectiveness, more comfort and fast diagnostics property. Soft computing is considered in biomedical field to analyze the image data properly and also to find out a connection among the diagnosis tools, treatment and forecasting of results. Medical data classification during diagnosis is the unavoidable problem, and this limitation can overcome by executing the techniques of soft computing.

Research performed in the field of intelligent systems comprises the development of certain techniques of intelligence computation in addition to application of such methods for real-time tasks. The methods of intelligent systems contain artificial intelligence (AI) algorithms which include machine learning, knowledge reasoning, etc. (Yan et al. 2015, Bhandari et al. 2020). A project named 'Programmable artificial cell evolution' targets to create a nanoscaled artificial protocol which can self-replicate as well as grow under controlled situations. Here, evolutionary modelling is executed to examine the dynamics of real and simulated protocell as well as the progression of photocellular networks. In order to develop stable self-replicating cell-membranes, evolution is executed by employing a genetic algorithm within the chips of microfluidic FPGA with the help of physical population mechanism on the chip. Additionally, computational modelling on progressive systems and embryonic development is becoming growing popularity in computational research. To imply artificial cell into real-time application, such techniques can offer a method for gene programming to facilitate the evolution of multicellular forms. However, a hybrid decision support system is designed using soft computing models which is resulted as a successful tool to detect the various stages of cervical cancer (Whitesides and Grzybowski 2002). In this reported work, hybridization is performed by considering a knowledge-grounded subnetwork section along with genetic algorithms where not only rough set theory is used but also Interactive Dichotomizer 3 algorithm is executed. The designed hybrid system shows enhanced performance with regard to classification outcome, training period and size of the network. This is an example of successful methodology for appropriate abstraction of logical rules and respective human interpretation of inference process.

However, swarm intelligent techniques are employed in various biomedical applications (Whitesides and Grzybowski 2002, Cai et al. 2007, O'Grady et al. 2008). A novel approach is found in medical image processing by introducing an algorithm named 'Stochastic Diffusion Search' (Ma et al. 2015). This algorithm can aid metastasis identification in the scanning of bones successfully. This algorithm is also implemented in a two-phase method of aorta identification in the images obtained from CT scan. Moreover, for aorta calcifications, swarm intelligent method is employed to obtain accurate localization (Alhakbani et al. 2016). However, in biomedical image processing, image segmentation shows a significant role. Additionally, for image segmentation, fuzzy c-means (FCM) clustering can be considered as a promising clustering algorithm. As FCM is sensitive to noise disruption, to overcome this limitation, a hybrid artificial fish swarm algorithm (HAFSA) is employed. The combination of artificial fish swarm algorithm (AFSA) and FCM results in the proposed algorithm, and it is utilized to obtain a superior outcome. Nevertheless, after introducing the mechanism of noise reduction and Metropolis principle, enhanced convergence rate has been found. For real-time application of the algorithm, magnetic resonance imaging (MRI), multidetector computed tomography (MDCT) as well as artificial grid graph are considered. The experimental outcomes confirm the enhanced anti-noise capacity with higher precision of the considered algorithm (Ma et al. 2015, Kim et al. 2015, Yan et al. 2015, Bhandari et al. 2020).

It is also observed that integrated magnetoresistive spintronic devices are used in various biomedical applications. In order to detect biomolecular recognition, biochip-based platforms are utilized. Moreover, for cell parting and cell counting, microfludic platforms are executed. Few neuroelectric devices are fabricated for brain acquisition, and magneto-cardiology devices are fabricated in order to obtain heart biomagnetic signals. Lab-on-a-chip platforms are found to use for distinct molecule actuation. Indeed, for single molecule understanding, magnetoresistive sensors are unified with the on-chip magnetic counterparts to yield nearly 1.0 ± 0.3 pN in order to perform magnetic bead actuation (Freitas et al. 2011, Hirohata et al. 2020). Such fabricated system is observed to use for characterization of single molecular level real-time bio-interaction. In addition, fabrication of hybrid devices combined with the flux guides and magnetoresistive element is used in the application of biomedical imaging. The usage of giant magnetoresistance sensor permits the magnetic fields' detection received from both the brain and heart of human even at a room temperature. Such magnetoresistive sensors can be decorated in nanoscaled range by labeling with nanosized magnetic particles, which makes an enhancement in the spatial resolution of the fabricated device. Herein, to count the target magnetic nanoparticles as well as the magnetically labelled cells, spin valve sensors are utilized.

In particular, the incorporation of fuzzy logic in DNA nanotechnology provides a significant remark in the field of biomedical application. The biological circuits found in natural living organism are functioned with molecules known as DNA and RNA not like the electronic transistors which are comprised of Boolean logic gates. The molecular logic gates and fabrication of molecular devices are the burning research topic in current research scenario. A self-assembled DNA complex has been designed which can sense multiple environmental signals to yield a fluorescent output. The developed logic gate can execute all the six Boolean logic gates operations along with a fuzzy logic gate operation having enhanced fidelity as well as reliability (Zadegan et al. 2015). The developed universal DNA logic gates can be applied to more complicated structures. In this aspect, the developed logic gate of NOR-DNA is implemented to a DNA-origami box. Such complicated self-assembled structures provide the possible controlling of multifunctional operation in order to achieve a novel smart biocalculator as well as nanomedical magnetic devices. The in-vivo calculation is found to perform successfully with this reported system with a potential efficiency toward cancer diagnostic as well as molecular level treatment. Certainly, such combined DNA logic gates are utilized to detect specific miRNA profiles, which can signify various types of cancers. The implementation of DNA origami box with logic gates permits unbolting of the lid for specific RNA input, which results in unveiling of a therapeutic agent, for example, enzyme, toxin or antibody confined in that respective origami box. While entering into the cell, the DNA origami box is sensitive to the cancerous cells. In addition, the sensor is programmed in such a way so that it can subsequently release the drug in order to destroy the tumor cells. This is a smart targeted procedure which responds only to the targeted cancerous cell so that it cannot induce any side-effects to the healthy tissues. The superparamagnetic property of nanoscaled material will enhance the efficiency of the smart drugs.

Meanwhile, these developments provide solutions to achieve smart biomedical tools which will solve various problems residing in the society.

3.5 CONCLUSION

It is observed from this systematic review that incorporation of SC techniques in the biomedical field can draw a connection between soft computing and biomedical nanotechnology directly. In the near future, it is possible to visualize a world where biomedical nanomagnetic devices are usually implanted in order to monitor the health as well as automatic repairing of damaged tissues by inserting such nanotherapeutic devices directly to the bloodstream. However, various subsections of artificial intelligence are found as useful techniques to implement during device fabrication in order to obtain smart devices to be used as a biomedical tool as well as smart drugs. In order to execute further research on SC techniques, following research gaps should be taken into consideration:

- Though fuzzy logic offers an effectual technique for linguistic knowledge illustration and operation, a real time application of such systematic and definitive technique is yet to be established and the exploration in this field will provide a new dimension for the researchers.
- In FL, when the input variables become larger with confounding interaction, it is challenging to analyze the consequences of the input variables on the resulted outcomes. In this aspect, a better engineering tool is required to modularize the applied model. Such limitations should be considered during the application of fuzzy logic to material science.
- It is understood that there are few inherent limitations of fuzzy logic because the procedure of membership functions generation is completely dependent on its context and may be affected by design predilections. However, this is considered as a lacuna of FL technique because, in real-time applications, it is better to use a systematic method having a well laid out pattern.
- There is still a smaller number of research studies in nanotechnology regarding incorporation of ANN techniques for better device fabrication. More research is required for better understanding to facilitate such networks more efficiently in real-time applications.
- Most of the spintronic devices and intelligence algorithms are yet found at a proof-of-concept level. Therefore, further in-depth research on potential implementation of SC techniques in device fabrication is essentially important.
- Systematic investigation on heterogeneity in Particle Swarm Optimization (PSO) algorithm is not observed yet, and hence, an understanding gap is found regarding the selection of exact parameter and topology in PSO.
- In case of a complicated self-assembly, effectual control mechanism is needed to be addressed, and requisition of extra potential characterizations is required. In nanorobot swarm systems, additional features are essential in order to simulate advanced nanorobotic systems.
- The training of neural networks is easy when there are moderate amounts of neurons. Such complicated structures for a certain pattern can be accomplished to limited errors; however, such networks are unable to provide exact response to novel patterns, whereas to utilize the neural networks appropriately, the structure should be simplest to execute the desired function.

Moreover, upgrading the existing fabricated devices by overcoming the limitations of SC techniques and development of novel smart devices will lead to a new pathway for better biomedical facilities.

REFERENCES

Abioduna, O. I., Jantanb, A., Omolara, A. E., et al. 2018. State-of-the-art in artificial neural network applications: Asurvey. *Heliyon* 4:e00938. https://sci-hub.do/10.1016/j.heliyon.2018.e00938.

Agrahari, R., Foroushani, A., Docking, T. R., et al. 2018.Applications of Bayesian network models in predicting types of hematological malignancies. *Scientific Reports* 8:6951.

Agrawal, S., Panda, R., and Dora, L. 2014. A study on fuzzy clustering for magnetic resonance brain image segmentation using soft computing approaches. *Applied Soft Computing* 24:522–533.

Alhakbani, H. A., and al-Rifaie, M. M. 2016. A swarm intelligence approach in undersampling majority class. *Swarm Intelligence* 9882:225.

Altameem, T., Zanaty, E. A., and Tolba, A. 2014. A new fuzzy C-means method for magnetic resonance image brain segmentation. *Connection Science* 27:305–321.

Attarzadeh, I., and Ow, S. H. 2010. A novel algorithmic cost estimation model based on soft computing technique. *Journal of Computer Science* 6:117–125.

Azimi, R., Yazdani-Chamzini, A., Fouladgar, M. M., et al. 2011. Ranking the strategies of mining sector through ANP and TOPSIS in a SWOT framework. *Journal of Business Economics and Management* 12:670–689.

Bhandari, A., Koppen, J., and Agzarian, M. 2020 Convolutional neural networks for braintumour segmentation. *Insights into Imaging* 11:77.

Bonsma E., Karunatillake N.C., Shipman, R., et al. 2003. Evolving greenfield passive optical networks. *BT Technology Journal* 21:44–4.

Cai, X., Cui, Z., Zeng, J., et al. 2007. Perceptive particle swarm optimization: A new learning method from birds seeking. *Computational and Ambient Intelligence* 4507:1130–1137.

Chandrasekaran, S., and Hougen, D. F. 2006. Swarm intelligence for cooperation of bio-nano robots using quorum sensing. *Bio Micro and Nanosystems Conference*, San Francisco, CA: 15–18. Doi: 10.1109/BMN.2006.330901.

Chandrasekaran, M., Muralidhar, M., Krishna, C. M., et al. 2010. Application of soft computing techniques in machining performance prediction and optimization: A literature review. *The International Journal of Advanced Manufacturing Technology* 46:445–464.

Fahim, H. Li, W. Javaid, S. et al., 2019. Fuzzy logic and bio-inspired firefly algorithm based routing scheme in intrabody nanonetworks. *Sensors* 19:5526.

Falvo, M.R., Clary, G.J., Taylor, R.M., et al. 1997. Bending and buckling of carbon nanotubes under large strain. *Nature* 389:582–584.

Freitas, P. P., Cardoso, F., Romao, V. C., et al. 2011. Spintronic platforms for biomedical applications. *Lab on a Chip* 12:546–57.

Hirohata, A., Yamada, K., Nakatani, Y., et al. 2020. Review on spintronics: Principles and device applications. *Journal of Magnetism and Magnetic Materials* 509:166711.

Ibrahim, D. 2016. An overview of soft computing. *Procedia Computer Science* 102: 34–38.

Kaur, P., Sharma, P., and Palmia, A. 2020. Fuzzy clustering-based image segmentation techniques used to segment magnetic resonance imaging/computed tomography scan brain tissues: Comparative analysis. *International Journal of Imaging Systems and Technology* 30:1294–1323.

Kim, K.B., Song, D. H., and Park, H. J. 2015 Computer vision based automatic extraction and thickness measurement of deep cervical flexor from ultrasonic images. *Computational and Mathematical Methods in Medicine* 2016: Article ID 5892051 (11 page).

Konwar, N., Bijan, D., and Debnath, P. 2019. Results on generalized intuitionistic fuzzy hypergroupoids. *Journal of Intelligent & Fuzzy Systems* 36:2571–2580.

Lee, J. H., Ahn, C. W., and An, J. 2013. An approach to self-assembling swarm robots using multitree genetic. *The Scientific World Journal* 2013:Article ID 593848 (10 pages).

Lee, D. S., Jung, H. Y., Lim, J. W., et al. 2000. Explosive gas recognition system using thick film sensor array and neural network. *Sensors and Actuators B* 71:90–98.

Lee, D. S., Kim, Y. T., Huh, J. S., et al. 2002. Fabrication and characteristics of SnO_2 gas sensor array for volatile organic compounds recognition. *Thin Solid Films* 416:271–278.

Liao, S., and Seeman, N.C. 2004. Translation of DNA signals into polymer assembly instructions. *Science* 306:2072–2074.

Ma, L., Li, Y. Fan, S. et al. 2015. A hybrid method for image segmentation based on artificial fish swarm algorithm and Fuzzy *c*-means clustering. *Computational and Mathematical Methods in Medicine* 2015: Article ID 120495 (10 page).

Nikravesh, M. 2008. Evolution of fuzzy logic: From intelligent systems and computationto human mind. *Forging New Frontiers: Fuzzy Pioneers* 12:207–214.

O'Grady, R., Christensen, A.L., and Dorigo, M. 2008. Autonomous reconfiguration in a self-assembling multi-robot system. *Ant Colony Optimization and Swarm Intelligence* 5217:259–266.

O'Grady, R., Groß, R., Christensen, A.L., et al. 2010. Self-assembly strategies in a group of autonomous mobile robots. *Autonomous Robots* 28:439–455.

Pezzella, F., Morganti, G., and Ciaschetti, G. 2008. A genetic algorithm for the flexible job-shop scheduling problem. *Computers & Operations Research* 35: 3202–3212.

Reddy, M. J., and Kumar, D.N. 2020. Evolutionary algorithms, swarm intelligence methods, and their applications in water resources engineering: A state-of-the-art review. *H_2Open Journal* 3: 135–188.

Ren, Q., and Cao, Q. 2006. Predictive model based on artificial neural net for purity of perovskite type $SrTiO_3$ nanocrystalline. *Transaction of Nonferrous Metals. Society of China* 16:865–868.

Shariati, S., Abedi, M., Saedi, A. et al., 2017. Critical factors of the application of nanotechnology in construction industry by using ANP technique under fuzzy intuitionistic environment. *Journal of Civil Engineering and Management* 23:914–925.

Shrivastava, S., and Dash, D. 2009. Applying nanotechnology to human health: Revolution in biomedical sciences. *Journal of Nanotechnology* 2009: Article ID 184702 (14 pages).

Suneetha, E., Karthikeyan, V., and Sujatha, K. 2020. Research on nano sensors on bio medical applications. *International Journal of Recent Technology and Engineering* 8: 5238–5245.

Vashist, S. K., Venkatesh A. G., KonstantinosMitsakakis, K., et al. 2012. Nanotechnology-based biosensors and diagnostics: Technology push versus industrial/healthcare requirements. *BioNano Science* 2:115–126.

Whitesides G.M., and Grzybowski B. 2002. Self-assembly at all scales. *Science* 295:2418–2421.

Yan, I., Zhu, R., Mi, L. et al. 2015 Segmentation of white blood cell from acute lymphoblastic leukemia images using dual-threshold method. *Computational and Mathematical Methods in Medicine*. 2016:1–12.

Yue, Z., and Jia, Y. 2013. An application of soft computing technique in group decision making under interval-valued intuitionistic fuzzy environment. *Applied Soft Computing* 13: 2490–2503.

Zadegan, R. M., Jepsen, M. D. E., Hildebrandt, L. L., et al. 2015. Construction of a fuzzy and boolean logic gates based on DNA. *Small* 11: 1811–1817.

Zarei, O., Fesanghary, M., Farshi, B., et al. 2009. Optimization of multi-pass face-milling via harmony search algorithm. *Journal Materials Processing Technology* 209:2386–2392.

4 A Mediative Fuzzy Logic-Based Approach to the Goal Programming Problem: In the Context of Multi-Objective Solid Transportation Problem

M. K. Sharma, Nitesh Dhiman, and Kamini
C.C.S. University

Vishnu Narayan Mishra
Indira Gandhi National Tribal University

CONTENTS

4.1 INTRODUCTION: BACKGROUND OF RELEVANT WORK

Uncertainty occurs in parameters while dealing with the distribution of goods from various origins to different destinations in transportation problems. Hitchcock introduced the classical transportation problem (Hitchcock 1941). Classical transportation problem is an unusual class of linear programming problem. It deals with transportation cost in connection with various constraints and it takes care of the demand of all the customers. Parameters involved in transportation problems contain different types of uncertainty and fuzziness. In 1962, Haley introduced a solid transportation problem in which uncertainty was present in the form of demand, supply, and conveyance (Haley 1962). The problems having only single conveyances with respect to the solid transportation problem is known as a classical transportation problem. Rather, it is to be noted that classical transportation problem is the unusual case of solid transportation problem, which depends upon the number of conveyances. Conventional theory is inefficient to deal with such types of uncertainties involved in such problems. In 1965, L. A. Zadeh provided the perception of fuzzy logic, which deals with such kinds of uncertainties by representing membership grade value to each point of the given set (Zadeh 1965). The problem becomes more imprecise when we have more than one uncertain factor involved in it; then in this case, the transport cost is not the only objective of transportation problem, but also it consists of more than one objective, which are conflicting in nature. Such kind of transportation problems is known as Multi-Objective Transportation Problems abbreviated as MOTP. There are many studies existing in the literature which handle the multi-objective problem in a better way as compared to other existing logics (Zimmermann 1978, Bit et al. 1999). In 1980, a goal programming problem was introduced in fuzzy environment by Narasimhan (1980). Wai el et al. also introduced goal programming in interactive environment for the transportation problem with multiple objectives (Wai el et al. 2006). In 2007, Zangiabadi and Maleki also gave a concept on goal programming dealing with fuzzy environment for transportation problem (Zangiabadi and Maleki 2007). Later in 2014, fractional-based goal programming has also been used to solve solid transportation problem with interval value cost (Radhakrishnan and Anukokila 2014). Later on, in 2019, Sharma et al. gave a novel approach on goal programming for transportation problem having multiple objectives (Sharma et al. 2019). Due to only the membership grade, the concept of fuzzy logic was inadequate to deal with such uncertainties which involve only the favorable cases.

Atanassov in 1986 dealt with a new type of logic which gives us a better result as compared to the traditional and fuzzy logic known as intuitionistic fuzzy logic (Atanassov 1986). In this logic, we can cover the uncertainty in more imprecise way by using the favorable and unfavorable cases simultaneously. Later, in 1989, Atanassov and Gargov have also introduced the interval-valued intuitionistic fuzzy set by using interval valued membership and nonmembership functions (Atanassov and Gargov 1989). In 2016, ranking method has also been used to deal intuitionistic fuzzy solid transportation problems (Aggarwal and Gupta 2016). Sometimes, the problem evolves when the data are collected from partial or absolute contradictory information and knowledge. Such kind of situation cannot be handled with the help of fuzzy as well as intuitionistic fuzzy logic. To handle such type of contradictory and noncontradictory information, Montiel et al. provide a mediated solution called mediative fuzzy logic (Montiel et al. 2005). Mediative fuzzy logic can be viewed as an extension of intuitionistic fuzzy logic in the view of Atanassov (1999). A lot of applications of mediative fuzzy logic have been studied so far, in population size (Montiel et al. 2009), in medicine (Iancu 2018, Dhiman and Sharma 2019), and in multiple criteria decision-making (Sharma and Dhiman 2019).

In this present chapter, we extended a comprehensive methodology in mediative environment that covers uncertainty and fuzziness with the help of favorable, unfavorable, and contradictory cases. There are many methodologies and techniques present in the existing literature to obtain an optimal solution for transportation problems with multiple objectives to obtain all the objectives simultaneously. In this proposed mathematical approach, we convert membership/nonmembership values of each parameter into a single mediative value with the help of mediative fuzzy logic. Then, to find the compromise optimal solution for multi-objective solid transportation problem with mediative fuzzy cost, we use a fuzzy goal programming approach with a deviational variable at minimum level.

4.2 BASIC CONCEPTS RELATED TO THE WORK

4.2.1 Fuzzy Set

A fuzzy set A defined on a universal set X is denoted as

$$A=\{(x,\ \mu_A(x)):\ x\in X\} \tag{4.1}$$

where $\mu_A(x)): X\rightarrow[0,\ 1]$ represents the membership value of A given in universal set X.

4.2.2 Intuitionistic Fuzzy Set

An intuitionistic fuzzy set B defined on a universal set X is denoted as

$$B=\{(x,\ \mu_B(x),\ \nu_B(x)):\ x\in X\} \tag{4.2}$$

where $\mu_B(x)$ and $\nu_B(x): X\rightarrow[0,\ 1]$ represents the membership value and nonmembership value of B, respectively, and $\pi_B(x)=1-[\mu_B(x)+\nu_B(x)]$ called the hesitation part of x in B, which lies between 0 and 1.

A trapezoidal intuitionistic fuzzy set is defined with following membership and nonmembership functions as

$$\mu_B(x)=\begin{cases}\dfrac{x-p}{q-p}\ \text{if}\ p<x<q\\ 1\ \text{if}\ q\le x\le r\\ \dfrac{s-x}{s-r}\ \text{if}\ r<x<s\end{cases}\quad \text{and}\,\nu_B(x)=\begin{cases}\dfrac{x-q}{q-p^*} & \text{if}\,p^*<x<q\\ 0 & \text{if}\ q\le x\le r\\ \dfrac{r-x}{s^*-r} & \text{if}\,r<x<s^*\end{cases}$$

where $p^*<p<q<r<s<s^*$ on a real line R.

Furthermore, let FS_μ and FS_ν represent traditional outputs of fuzzy system using membership and nonmembership grade, respectively; then, the overall output of a system based on intuitionistic fuzzy logic is represented by a linear relation given by (Castillo and Melin 2003) as

$$y = (1-\pi)FS_\mu + \pi FS_\nu \tag{4.3}$$

If $\pi = 0$, then it converts into the output for the traditional fuzzy-based system, but if $\pi \neq 0$, we have the output for the intuitionistic fuzzy system.

4.2.3 Mediative Fuzzy Logic

The theory of mediative fuzzy set initialized with a general overview of contradiction set. A contradictory fuzzy set C in X is defined by

$$\zeta_C(x) = \min(\mu_C(x), \nu_C(x)) \tag{4.4}$$

Based on this, Montiel et al. (2009) suggested the following three expressions as

$$\text{MFS} = \left(1-\pi-\frac{\zeta}{2}\right)FS_\mu + \left(\pi+\frac{\zeta}{2}\right)FS_\nu \tag{4.5}$$

$$\text{MFS} = \min\left[\left((1-\pi)FS_\mu + \pi FS_\nu\right), \left(1-\frac{\zeta}{2}\right)\right] \tag{4.6}$$

$$\text{MFS} = \left((1-\pi)FS_\mu + \pi FS_\nu\right)\left(1-\frac{\zeta}{2}\right) \tag{4.7}$$

From these three equations, we can calculate the output of the mediative fuzzy logic-based system. If $\zeta = 0$, then it converts into the output of the system based on intuitionistic fuzzy logic, but if $\zeta \neq 0$, we have the output for the mediative fuzzy system.

4.2.4 Transportation Problem

Let there be m sources and n destinations. Let a_i be available quantity of the products which are shipped from i^{th} source, and b_j be demand of the product of jth destination. c_{ij} be the cost of the products which are delivered from the ith source to the jth destination. x_{ij} be the quantity of the products with cost c_{ij}, delivered from the ith source to the jth destination. Transportation problem can be stated as linear programming problem in the following manner: -

$$\text{Min } Z(x) = \sum_{i=1}^{m}\sum_{j=1}^{n} c_{ij}x_{ij}$$

Subject to constraints:

$$\sum_{i=1}^{m} x_{ij} = b_j, \ \forall \text{ j}$$

$$\sum_{j=1}^{n} x_{ij} = a_i, \ \forall \ i$$

$$\sum_{i=1}^{m} a_i = \sum_{j=1}^{n} b_j$$

where $i = 1,2,\ldots,m$ & $j = 1,2,\ldots,n$.

4.2.5 Solid Transportation Problem

Let there be m number of sources and n number of destinations. Let a_{jk} be available quantity of products which shipped to j^{th} destination by k^{th} conveyance, b_{ik} be the demand of the products which delivered from i^{th} source by k^{th} conveyance, and e_{ij} be the quantity of products which delivered from i^{th} source to j^{th} destination. Let x_{ijk} be the number of products delivered from i^{th} source to j^{th} destination with k^{th} conveyance and c_{ijk} be the cost per unit of products which delivered from i^{th} source to j^{th} destination by k^{th} conveyance, where $i = 1,2,\ldots,m$, $j = 1,2,\ldots,n$ and $k = 1,2,\ldots,K$.

Then, the mathematical formulation of Solid Transportation problem can be defined in the following manner:

$$\text{Min } Z(x) = \sum_{i=1}^{m}\sum_{j=1}^{n}\sum_{k=1}^{K} c_{ijk} x_{ijk}$$

subject to constraints:

$$\sum_{i=1}^{m} x_{ijk} = a_{jk},$$

$$\sum_{j=1}^{n} x_{ijk} = b_{ik},$$

$$\sum_{k=1}^{K} x_{ijk} = e_{ij},$$

where $\sum_{j=1}^{n} a_{jk} = \sum_{i=1}^{m} b_{ik}, \ \sum_{j=1}^{n} e_{ij} = \sum_{k=1}^{K} b_{ik}, \ \sum_{i=1}^{m} e_{ij} = \sum_{k=1}^{K} a_{jk},$

$$\sum_{j=1}^{n}\sum_{k=1}^{K} a_{jk} = \sum_{i=1}^{m}\sum_{k=1}^{K} b_{ik} = \sum_{i=1}^{m}\sum_{j=1}^{n} e_{ij},$$

$x_{ijk} \geq 0$, where $i = 1, 2\ldots m, j = 1, 2\ldots.n$ and $k = 1,2\ldots K$.

4.2.6 Solid Transportation Problem with Multiple Objectives

Let there be a problem with m sources, n destination, K number of conveyance and D number of objectives. Let a_{jk} be available quantity of products delivered to j^{th} destination by k^{th} conveyance, b_{ik} be the demand of the products delivered from i^{th} source by k^{th} conveyance, and e_{ij} be the quantity of products delivered from i^{th} source to j^{th} destination. Let x_{ijk} be the no. of products delivered from i^{th} source to j^{th} destination by k^{th} conveyance, and c_{ijk} be the cost per unit of products delivered from i^{th} source to j^{th} destination by k^{th} conveyance, where $i = 1,2\ldots.m, j = 1,2\ldots.n$ and $k = 1,2\ldots.K$. Then, the mathematical representation of solid transportation problem with multiple objectives is as follows:

$$\text{Min } Z_d = \sum_{i=1}^{m}\sum_{j=1}^{n}\sum_{k=1}^{K} c_{ijk}^{d} x_{ijk} \text{where } d = 1,\ 2\ldots.\ D.$$

subject to the constraints:

$$\sum_{i=1}^{m} x_{ijk} = a_{jk},$$

$$\sum_{j=1}^{n} x_{ijk} = b_{ik},$$

$$\sum_{k=1}^{K} x_{ijk} = e_{ij},$$

$$\text{where } \sum_{j=1}^{n} a_{jk} = \sum_{i=1}^{m} b_{ik},\ \sum_{j=1}^{n} e_{ij} = \sum_{k=1}^{K} b_{ik},\ \sum_{i=1}^{m} e_{ij} = \sum_{k=1}^{K} a_{jk},$$

$$\sum_{j=1}^{n}\sum_{k=1}^{K} a_{jk} = \sum_{i=1}^{m}\sum_{k=1}^{K} b_{ik} = \sum_{i=1}^{m}\sum_{j=1}^{n} e_{ij},$$

$x_{ijk} \geq 0$, where $i = 1,2,\ldots.,m, j = 1,2,\ldots.,n$ and $k = 1,2,\ldots.,K$.

4.2.7 Goal Programming

Goal programming problem is a problem in which decision maker sets a desired level for the respective objectives. Firstly, goal programming problem was discussed by Charnes and Cooper. Let Z_d be the objectives and $\bar{Z}_d$ be the aspiration level of the objective Z_d for each d, decided by the ruling maker. In the goal programming problem, ruling maker wants to minimize the deviation between objective Z_d and its aspiration level $\bar{Z}_d$. There are two types of deviational variable arises in the problem: when $Z_d \geq \bar{Z}_d$ then overachievement deviational variable $d_d^+ = \frac{1}{2}\left\{(Z_d - \bar{Z}_d) + \left|Z_d - \bar{Z}_d\right|\right\}$ arises and when $Z_d \leq \bar{Z}_d$ then underachievement deviational variable $d_d^- = \frac{1}{2}\left\{(\bar{Z}_k - Z_k) + \left|\bar{Z}_k - Z_k\right|\right\}$ arises.

Then the mathematical formulation of goal programming:

$$\text{Min} \sum_{d=1}^{D} (d_d^+ - d_d^-)$$

$$\text{Subject to} \sum_{i=1}^{m} \sum_{j=1}^{n} c_{ij} x_{ij} - \bar{Z}_d = d_d^+ - d_d^-,$$

$$\sum_{i=1}^{m} x_{ij} = b_j,$$

$$\sum_{j=1}^{n} x_{ij} = a_i,$$

$$d_d^+ d_d^- = 0, d_d^+ - d_d^- \geq 0, \; d = 1,2 \ldots . D.$$

where Z_d and $\bar{Z}_d$ are d^{th} objective function and aspiration level of dth objective function, respectively.

4.2.8 Fuzzy Goal Programming (FGP)

In fuzzy decision-making surroundings, goals of the objectives are fuzzy. The concept of fuzzy in goal programming problem was introduced by Narasimhan in 1980. In FGP, accomplishment of a fuzzy goal to its aspiration level means accomplishment of the membership function of the goal to its highest membership grade. Since the highest membership grade of the membership function is 1, now, decision maker wants to accomplish the highest membership grade of the membership function of the goal by minimizing the below deviational variables.

Then the mathematical formulation of FGP is as follows:

$$\text{Min } \delta,$$

Subject to

$$\mu_d(Z_d(x)) - d_d^+ + d_d^- = 1,$$

$$\sum_{i=1}^{m} x_{ij} = b_j,$$

$$\sum_{j=1}^{n} x_{ij} = a_i,\ \delta \geq d_d^-,$$

$$d_d^+ d_d^- = 0,\ \delta \geq 0, \delta \leq 1,$$

$x_{ij} \geq 0$, Where $i = 1, 2,\ldots,m$, $j = 1,2,\ldots, n$ and $d = 1,2,\ldots, D$.

4.3 MATHEMATICAL STRUCTURE FORMULATION FOR THE PROPOSED ALGORITHM (MFMSTP)

The framework of the proposed work on Mediative fuzzy logic to multi-objective solid transportation problem consists of certain steps shown as follows:

Step 1 – First, consider a solid transportation problem having multi-objective with intuitionistic fuzzy cost as shown in Figure 4.1.

Step 2 – Convert MOSTP into crisp mode by using mediative function based on mediative fuzzy logic and the mediate value for intuitionistic fuzzy set as:

$$y = \left(1 - \pi - \frac{\zeta}{2}\right) y^1 + \left(\pi + \frac{\zeta}{2}\right) y^2 \tag{4.8}$$

where $\mu = \dfrac{U - y^1}{U - L}$, $\nu = \dfrac{y^2 - L'}{U' - L'}$ and U, U' are the upper bounds of membership and nonmembership functions, and L, L' are the lower bounds of membership and nonmembership functions.

Where, y^1 and y^2 are the defuzzified values of membership and nonmembership grade, respectively.

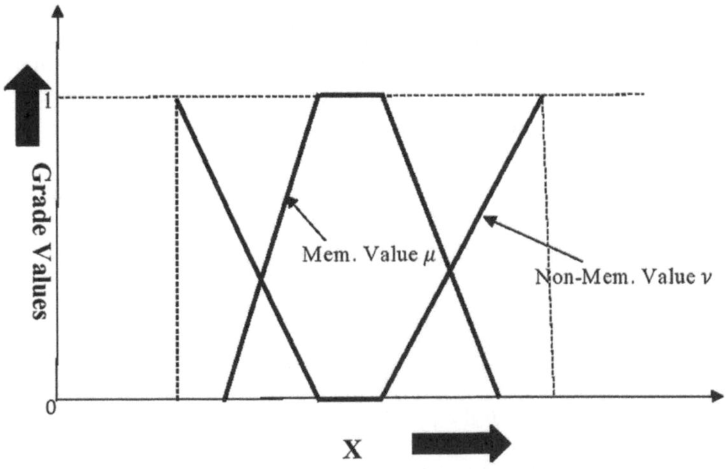

FIGURE 4.1 Geometrical representation of trapezoidal intuitionistic fuzzy set.

TABLE 4.1
Payoff Matrix for All Objectives

	X_i
Z_i	

Step 3 – Check whether the given problem is balanced, otherwise convert into balanced problem by adding a dummy row or column with zero cost.

Step 4 – Now, solve the problem by Vogel's approximation method taken single objective at a time for each j, respectively. Then, we get basic feasible solutions for all objectives of the problem.

Step 5 – Now, apply the method proposed to intuitionistic fuzzy cost for the optimal solution of the respective objectives. The method will be as follows:

i. After finding the solution by Vogel's approximation method, for each $j = 1$, check if allocation of c_{i1k} exceeds b_{ik} or not. If not, then find the solution for each j by Vogel's approximation method.
ii. If it exceeds by θ units, then make a loop starting with c_{i1k} and move horizontally and vertically with allocation filling up cells. Now, assign $+\theta$ and $-\theta$ sign alternatively at each corner starting $-\theta$ at c_{i1k}. All cells must be filled up with minimum θ allocation.
iii. Now, if any entry at c_{i1k} with +ve sign in loop is not filled up, then least value at b_{ik} must be θ. If it is filled up with β value, then least value at b_{ik} must be $\theta+\ \beta$. Do the same process till all three constraints are not satisfied.

Step 6 – Now, make the payoff matrix for all objectives as shown in the Table 4.1, where X_i is the solution obtained by step-5 for Z_i, respectively.

Step 7 – Determine the lower bound L_d and upper bound U_d for the objective function given in the problem.

Step 8 – Now, to get a compromise optimal solution for all objectives simultaneously, build the model for given problem and then solve by using Lingo software.

4.4 NUMERICAL EXAMPLE

Step 1 – Consider a solid transportation problem having multiple objectives with trapezoidal intuitionistic fuzzy cost as shown in Table 4.2.

Step 2 – Mediate the cost by the following functions of the problem:

$$y = \left(1 - \pi - \frac{\zeta}{2}\right)y^1 + \left(\pi + \frac{\zeta}{2}\right)y^2$$

where $\mu = \dfrac{U - y^1}{U - L}$, $v = \dfrac{y^2 - L'}{U' - L'}$ and U, U' are the upper bounds of membership and nonmembership functions and L, L' are the lower bounds of membership and nonmembership functions.

TABLE 4.2
Multi-Objective Solid Transportation Problem with Trapezoidal Intuitionistic Fuzzy Cost

	D_1	D_2	D_3	Supply
O_1	$e_{11} = 10$ $\tilde{c}_{111}\tilde{c}_{112}\tilde{c}_{113}$	$e_{12} = 6 = 6$ $\tilde{c}_{121}\tilde{c}_{122}\tilde{c}_{123}$	$e_{13} = 10$ $\tilde{c}_{131}\tilde{c}_{132}\tilde{c}_{133}$	$b_{11}b_{12}b_{13}$ 8, 10, 8
O_2	$e_{21} = 12$ $\tilde{c}_{211}\tilde{c}_{212}\tilde{c}_{213}$	$e_{22} = 9$ $\tilde{c}_{221}\tilde{c}_{222}\tilde{c}_{223}$	$e_{23} = 14$ $\tilde{c}_{231}\tilde{c}_{232}\tilde{c}_{233}$	$b_{21}b_{22}b_{23}$ 10, 12, 13
O_3	$e_{31} = 12$ $\tilde{c}_{311}\tilde{c}_{312}\tilde{c}_{313}$	$e_{32} = 9$ $\tilde{c}_{321}\tilde{c}_{322}\tilde{c}_{323}$	$e_{33} = 11$ $\tilde{c}_{331}\tilde{c}_{332}\tilde{c}_{333}$	$b_{31}b_{32}b_{33}$ 13, 9, 10
Demand	$a_{11}a_{12}a_{13}$ 10, 14, 10	$a_{21}a_{22}a_{23}$ 6, 10, 8	$a_{31}a_{32}a_{33}$ 15, 7, 13	

For both the objectives, i.e., for first and second objective, mediative fuzzy cost is represented in the form of trapezoidal intuitionistic fuzzy number with their respective mediative values, as shown in Tables 4.3 and 4.4.

Step 3 – Since

$$\sum_{j=1}^{n} a_{jk} = \sum_{i=1}^{m} b_{ik} = 31, \sum_{j=1}^{n} e_{ij} = \sum_{k=1}^{k} b_{ik} = 26, \sum_{j=1}^{m} e_{ij} = \sum_{k=1}^{K} a_{jk} = 34,$$

$$\sum_{j=1}^{n}\sum_{k=1}^{K} a_{jk} = \sum_{i=1}^{m}\sum_{k=1}^{K} b_{ik} = \sum_{i=1}^{m}\sum_{j=1}^{n} e_{ij} = 93.$$

Hence, the problem is a balanced problem.

Step 4 & 5 – Now by applying the proposed approach, we get the solution of both the objectives as follows:

For first objective solution X_1 is as follows: $x_{111} = 0$, $x_{112} = 10$, $x_{113} = 0$, $x_{211} = 0$, $x_{212} = 4$, $x_{213} = 8$, $x_{311} = 10$, $x_{312} = 0$, $x_{313} = 2$, $x_{121} = 0$, $x_{122} = 0$, $x_{123} = 6$, $x_{221} = 3$, $x_{222} = 6$, $x_{223} = 0$, $x_{321} = 3$, $x_{322} = 4$, $x_{323} = 2$, $x_{131} = 8$, $x_{132} = 0$, $x_{133} = 2$, $x_{231} = 7$, $x_{232} = 2$, $x_{233} = 5$,. $x_{331} = 0, x_{332} = 5, x_{333} = 6$.

At X_1, $Z_1 = 624.58$, $Z_2 = 743.78$.

For second objective solution X_2 is as follows: $x_{111} = 8$, $x_{112} = 2$, $x_{113} = 0$, $x_{211} = 2$, $x_{212} = 10$, $x_{213} = 0$, $x_{311} = 0$, $x_{312} = 0$, $x_{313} = 10$, $x_{121} = 0$, $x_{122} = 5$, $x_{123} = 1$, $x_{221} = 0$, $x_{222} = 2$, $x_{223} = 7$, $x_{321} = 6$, $x_{322} = 3$, $x_{323} = 0$, $x_{131} = 0$, $x_{132} = 3$, $x_{133} = 7$, $x_{231} = 8$, $x_{232} = 0$, $x_{233} = 6$, $x_{331} = 7, x_{332} = 4, x_{333} = 0$.

At X_2, $Z_1 = 658.41$, $Z_2 = 714.34$.

Step 6 – We have the payoff matrix as shown in Table 4.5.

Step 7 – Now, lower bound and upper bound of 1st objective $L_1 = 624.58$, $U_1 = 658.41$; and lower and upper bound of 2nd objective $L_2 = 714.34$, $U_2 = 743.78$.

TABLE 4.3
Trapezoidal Intuitionistic Fuzzy Costs and Their Mediative Values for the First Objective

Trapezoidal Intuitionistic Fuzzy Cost	Mediative Values
< (5, 6, 10, 13) (4, 6, 10, 14) >	7.80
< (4, 5, 7, 9) (3, 5, 7, 10) >	5.72
< (6, 8, 10, 12) (5, 8, 10, 13) >	8.961
< (8, 9, 11, 13) (7, 9, 11, 14) >	10.51
< (3, 5, 6, 9) (2, 5, 6, 10) >	4.72
< (2, 3, 7, 8) (1, 3, 7, 9) >	5
< (5, 6, 7, 9) (4, 6, 7, 10) >	6.67
< (2, 4, 6, 8) (1, 4, 6, 9) >	5
< (3, 4, 5, 7) (2, 4, 5, 8) >	4.5
< (2, 4, 6, 8) (1, 4, 6, 9) >	5
< (8, 9, 11, 13) (7, 9, 11, 14) >	10.51
< (2, 4, 6, 8) (1, 4, 6, 9) >	5
< (3, 5, 6, 9) (2, 5, 6, 10) >	4.72
< (5, 6, 10, 13) (4, 6, 10, 14) >	7.80
< (7, 9, 10, 12) (6, 9, 10, 13) >	9.5
< (3, 5, 6, 8) (2, 5, 6, 9) >	5.38
< (10, 11, 12, 14) (9, 11, 12, 15) >	11.75
< (6, 8, 10, 12) (5, 8, 10, 13) >	8.961
< (10, 11, 12, 14) (9, 11, 12, 15) >	11.75
< (8, 9, 11, 13) (7, 9, 11, 14) >	10.51
< (6, 8, 10, 12) (5, 8, 10, 13) >	8.961
< (3, 4, 5, 7) (2, 4, 5, 8) >	4.5
< (5, 6, 7, 9) (4, 6, 7, 10) >	6.67
< (2, 4, 6, 8) (1, 4, 6, 9) >	5
< (6, 8, 10, 12) (5, 8, 10, 13) >	8.961
< (5, 6, 10, 13) (4, 6, 10, 14) >	7.80
< (5, 6, 7, 9) (4, 6, 7, 10) >	6.67

Step 8 – Proposed model of the problem: -

$$\text{Min } A,$$

$$\mu_1(z_1(x)) + (d_1^- - d_1^+)(U_1 - L_1) = 1, \text{where } \mu_d\left(z_d(x)\right)$$

$$= \frac{U_d - Z_d}{U_d - L_d} \text{ if } L_d \le Z_d \le U_d, d = 1,2.$$

$$Z_1 - (d_1^- - d_1^+)(U_1 - L_1) = L_1$$

TABLE 4.4
Trapezoidal Intuitionistic Fuzzy Costs and Their Mediative Values for the Second Objective

Trapezoidal Intuitionistic Fuzzy Cost	Mediative Values
< (5, 6, 10, 13) (4, 6, 10, 14) >	7.80
< (10, 11, 12, 14) (8, 11, 12, 16) >	11.70
< (8, 9, 11, 13) (6, 9, 11, 15) >	10.46
< (8, 9, 11, 13) (6, 9, 11, 15) >	10.46
< (6, 8, 10, 12) (5, 8, 10, 13) >	8.961
< (4, 6, 7, 9) (3, 6, 7, 10) >	6.5
< (4, 6, 7, 9) (3, 6, 7, 10) >	6.5
< (6, 8, 10, 12) (5, 8, 10, 13) >	8.961
< (2, 4, 6, 8) (1, 4, 6, 9) >	5
< (3, 5, 6, 8) (2, 5, 6, 9) >	5.38
< (6, 8, 10, 12) (5, 8, 10, 13) >	8.961
< (4, 6, 7, 9) (2, 6, 7, 11) >	6
< (7, 9, 10, 12) (6, 9, 10, 13) >	9.5
< (10, 11, 12, 14) (8, 11, 12, 16) >	11.70
< (8, 9, 11, 13) (6, 9, 11, 15) >	10.46
< (4, 6, 7, 9) (2, 6, 7, 11) >	6
< (8, 9, 11, 13) (6, 9, 11, 15) >	10.46
< (7, 9, 10, 12) (6, 9, 10, 13) >	9.5
< (5, 6, 10, 13) (4, 6, 10, 14) >	7.80
< (4, 6, 7, 9) (3, 6, 7, 10) >	6.5
< (6, 8, 10, 12) (5, 8, 10, 13) >	8.961
< (3, 5, 6, 8) (2, 5, 6, 9) >	5.38
< (5, 6, 10, 13) (4, 6, 10, 14) >	7.80
< (7, 9, 10, 12) (6, 9, 10, 13) >	9.5
< (3, 5, 6, 9) (2, 5, 6, 10) >	4.72
< (2, 4, 6, 8) (1, 4, 6, 9) >	5
< (3, 5, 6, 9) (2, 5, 6, 10) >	4.72

TABLE 4.5
Payoff Matrix

	X_1	X_2
Z_1	624.58	658.41
Z_2	743.78	714.34

where $L_1 = 624.58$, $U_1 = 658.41$ and $U_1 - L_1 = 33.83$;
and $\mu_2(z_2(x)) + (d_2^- - d_2^+)(U_2 - L_2) = 1$

$$Z_2 - (d_2^- - d_2^+)(U_2 - L_2) = L_2$$

where $L_2 = 714.34$, $U_2 = 743.78$ and $U_2 - L_2 = 29.44$;
Constraints are given as follows:

$$x_{111} + x_{211} + x_{311} = 10, x_{112} + x_{212} + x_{312} = 14, x_{113} + x_{213} + x_{313} = 10$$

$$x_{121} + x_{221} + x_{321} = 6, x_{122} + x_{222} + x_{322} = 10, x_{123} + x_{223} + x_{323} = 8$$

$$x_{131} + x_{231} + x_{331} = 15, x_{132} + x_{232} + x_{332} = 7, x_{133} + x_{233} + x_{333} = 13$$

$$x_{111} + x_{121} + x_{131} = 8, x_{112} + x_{122} + x_{132} = 10, x_{113} + x_{123} + x_{133} = 8$$

$$x_{211} + x_{221} + x_{231} = 10, x_{212} + x_{222} + x_{232} = 12, x_{213} + x_{223} + x_{233} = 13$$

$$x_{311} + x_{321} + x_{331} = 13, x_{312} + x_{322} + x_{332} = 9, x_{313} + x_{323} + x_{333} = 10$$

$$x_{111} + x_{112} + x_{113} = 10, x_{211} + x_{212} + x_{213} = 12, x_{311} + x_{312} + x_{313} = 12$$

$$x_{121} + x_{122} + x_{123} = 6, x_{221} + x_{222} + x_{223} = 9, x_{321} + x_{322} + x_{323} = 9$$

$$x_{131} + x_{132} + x_{133} = 10, x_{231} + x_{232} + x_{233} = 14, x_{331} + x_{332} + x_{333} = 11$$

$A \geq d_1^-$, $A \geq d_2^-$ and $A \geq 0$ and $A \leq 1$.

$d_i^-.d_i^+ = 0$, where i = 1, 2, x_{ijk}, $d_{\bar{d}}$, $d_d^+ \geq 0$, where i = 1, 2...m, j = 1,2...n, k = 1, 2... K and d = 1, 2.

Then compromise solution for both objectives of problem obtained by Lingo 18:

$x_{111} = 8$, $x_{112} = 2$, $x_{113} = 0$, $x_{211} = 0$, $x_{212} = 4.463769$, $x_{213} = 7.536231$, $x_{311} = 2$, $x_{312} = 7.536231$, $x_{313} = 2.463769$, $x_{121} = 0$, $x_{122} = 2.881658$, $x_{123} = 3.118342$, $x_{221} = 0$, $x_{222} = 6.772496$, $x_{223} = 2.227504$, $x_{321} = 6$, $x_{322} = 0.3458455$, $x_{323} = 2.654154$, $x_{131} = 0$, $x_{132} = 5.118342$, $x_{133} = 4.881658$, $x_{231} = 10$, $x_{232} = 0.7637348$, $x_{233} = 3.236265$, $x_{331} = 5$, $x_{332} = 1.117923$, $x_{333} = 4.882077$, then $Z_1 = 624.58$, $Z_2 = 714.34$.

Since the obtained solution of both objectives is the min value in range of the solution, the obtained compromise solution is the best solution for the problem.

4.5 CONCLUSION

In this work, we introduced a new approach to handle the transportation problem for multiple objects with a trapezoidal intuitionistic fuzzy cost. The foremost objectives of the proposed mediative fuzzy logic approach are to optimize all objectives simultaneously. During this work, we have taken a solid transportation problem which involves three constraints (i.e. Demand, supply, and Conveyance) taken in Table 4.2. Initially, in the place of defuzzification of the parameters, the cost is handled by using mediative fuzzy logic. We converted membership and nonmembership grades of each parameter into a single mediated value. The values of the parameters are shown into a single mediated value (as shown in Tables 4.3 and 4.4). This theory illustrates the appropriate use of contradictory factors involved in the multi-objective transportation problem. The proposed mediative approach gave a more optimal solution than the solution obtained by defuzzified cost by ranking function. Traditional fuzzy logic and intuitionistic fuzzy logic are not able to deal with these situations, where the contradiction exists in expert knowledge. By obtaining the solution for each objective, we found the payoff matrix as shown in Table 4.5. From Table 4.5, we can easily observe that the obtained solution of both the objectives is the minimum value in range of the solution. Hence, the obtained compromise solution is the best solution for the problem as gained by the previous methods.

ACKNOWLEDGMENTS

The second author is grateful to the UGC (University Grant Commission) for economical encouragement.

REFERENCES

Aggarwal, S. and C. Gupta. 2016. Solving intuitionistic fuzzy solid transportation problem via new ranking method based on signed distance. *Int. J. Uncertain. Fuzziness Knowl.-Based Syst.* 24: 483–501.

Atanassov, K.T. 1999. *Intuitionistic Fuzzy Sets: Theory and Applications.* Springer. Heidelberg.

Atanassov, K.T. 1986. Intuitionistic fuzzy set. *Fuzzy Sets Syst.* 20: 87–97.

Atanassov, K.T., and G. Gargov. 1989. Interval-valued intuitionistic fuzzy sets. *Fuzzy Sets Syst.* 31: 343–349.

Bit, A.K., M.P. Biswal, and S.S. Alam. 1999. Multi-objective transportation problem with interval cost source and destination parameters. *Eur. J. Oper.* 117: 100–112.

Castillo, O. and P. Melin. 2003. A New method for fuzzy inference in intuitionistic fuzzy systems. *22nd International Conference of the North American Fuzzy Information Processing Society*, NAFIPS, Chicago.

Dhiman, N., and M.K. Sharma. 2019. Mediative Sugeno's-TSK fuzzy logic based screening analysis to diagnosis of heart disease. *Appl. Math.* 10:448–467.

Haley, K. 1962. The solid transportation problem. *Oper. Res.* 10: 448–463.

Hitchcock, F.L. 1941. The distribution of product from several sources to numerous localities. *MIT J. Math. Phys.* 20: 224–230.

Iancu, I. 2018. Heart disease diagnosis based on mediative fuzzy logic. *Artif. Intell Med.* 89: 51–60.

Montiel, O., O. Castillo, P. Melin, D.A. Rodríguez, and R. Sepúlveda. 2005. Reducing the cycling problem in evolutionary algorithms. In: *Proceedings of ICAI-2005*, 426–432, Las Vegas.

Montiel, O., O. Castillo, P. Melin, and M. Sepulveda. 2009. Mediative fuzzy logic for controlling population size in evolutionary algorithms. *Intell. Inf. Manag. 1*: 108–119.

Narasimhan, R. 1980. Goal programming in fuzzy environment. *Decis, Sci.* 111: 325–336.

Radhakrishnan, B., and P. Anukokila. 2014. Fractional goal programming for fuzzy solid transportation problem with interval cost. *Fuzzy Inf. Eng.* 6: 359–377.
Sharma, M.K. and N. Dhiman. 2019. Mediative multi-criteria decision support system for various alternatives based on fuzzy logic. *IJRTE* 8: 7940–7946.
Sharma, M.K., Kamini and N. Dhiman. 2019. A new goal programming approach for multi-objective solid transportation problem with interval-valued intuitionistic fuzzy logic. *JIAPS.* 23(3): 251–268.
Waiel, F.A.E.W., and S.M. Lee. 2006. Interactive fuzzy goal programming for multi-objective transportation problems. *Omega.* 34: 158–166.
Zadeh, L.A. 1965. Fuzzy sets. *Inf. Control.* 8: 338–353.
Zangiabadi, M. and H.R. Maleki. 2007. Fuzzy goal programming for multi-objective transportation problems. *J. Appl. Math Comput.* 24: 449–469.
Zimmermann, H.J. 1978. Fuzzy programming and linear programming with several objective Functions. *Fuzzy Sets Syst.* 1: 45–55.

5 H-U-R Stability Results of Mixed-Type Additive-Quadratic Functional Equation in Fuzzy β-Normed Spaces by Two Different Approaches

K. Tamilvanan
Government Arts College for Men

K. Loganathan
Erode Arts & Science College

N. Revathi
Periyar University PG Extension Centre

CONTENTS

5.1 INTRODUCTION

The concept of a fuzzy normed linear space was presented by C. Felbin (1992). He demonstrated that in a finite-dimensional fuzzy normed linear space, fuzzy norms are equivalent up to fuzzy proportionality. Finite-dimensional fuzzy subspaces of a fuzzy normed linear space ended up being fundamentally complete fuzzy normed linear spaces. The notions of fuzzy normed spaces were investigated by many authors (Bag and Samanta 2003, Cheng and Moderson 1994).

In the event that a mathematical equation admits a unique solution, we may state that the equation is stable. The primary stability issue concerning group homomorphisms was raised by Ulam (1960) and affirmatively derived by Hyers (1941). Applications of stability theory of functional equations for the evidence of new fixed-point theorems with applications may be found in the literature (Isac and Rassias 1996).

Functional equations are used in modelling several real-world problems (Cadariu and Radu 2008, Lowen 1976, Park et al. 2020). For such applications in physics, engineering, mechanics, biology, etc., we refer to (Alanazi et al. 2020, Lee et al. 2012, Tamilvanan et al. 2020a, 2020b, Park 2011, Radu 2003) where nonlinear functional equations have been used. For more relevant information about such functional equations, we suggest (Cadariu and Radu 2004, Kaleva and Seikkala 1984, Katsaras 1981, Katsaras and Liu 1977, Tamilvanan et al. 2020, Yang et al. 2015) and the references cited therein.

The following functional equations

$$\varphi(s+t)=\varphi(s)+\varphi(v) \tag{5.1}$$

$$\varphi(s+t)+\varphi(s-t)=2\varphi(s)+2\varphi(t) \tag{5.2}$$

are called the Cauchy additive and the quadratic functional equations. Both equations have the solutions $\varphi(s)=c\ s$ and $\varphi(s)=c\ s^2$, where c is an arbitrary constant. Each solution of the functional equations (5.1) and (5.2) are called additive and quadratic.

Definition 5.1

Assume that A is a real vector space. A mapping $N_\beta : A\times\mathbb{R}\to[0,1]$ is called a fuzzy β-norm on A with $0<\beta\leq 1$ if for all $v,w\in A$ and $p,q\in\mathbb{R}$,

(F1) $N_\beta(v,q)=0$ for all $q\leq 0$.
(F2) $v=0$ if $N_\beta(v,q)=1$ for all $q>0$;
(F3) $N_\beta(cv,q)=N_\beta(v,\frac{q}{|c|^\beta}$ if $c\neq 0$;
(F4) $N_\beta(v+w,p+q)\geq\min\{N_\beta(v,p),N_\beta(w,q)\}$;
(F5) $N_\beta(v,\cdot)$ is a nondecreasing function of $\mathbb{R}$ with

$$\lim_{q\to\infty} N_\beta(v,q)=1$$

(F6) for $v\neq 0$, $N_\beta(v,\cdot)$ is continuous on $\mathbb{R}$.

Then, (A,N_β) is named a fuzzy β-normed vector space.

In this work, the authors newly introduce the additive-quadratic functional eq. as

$$\varphi\left(\sum_{a=1}^{m} v_a\right)+\varphi\left(-v_b+\sum_{\substack{a=1;\\ a\neq b}}^{m} v_i\right)=(m-3)\sum_{a=1}^{m}\varphi(v_a+v_b)$$

$$-\left(m^2-5m+4\right)\sum_{a=1}^{m}\left[\frac{\varphi(v_a)-\varphi(-v_a)}{2}\right]$$

$$+\left(-m^2+5m-2\right)\sum_{a=1}^{m}\left[\frac{\varphi(v_a)+\varphi(-v_a)}{2}\right] \quad (5.3)$$

where $m \geq 3$ and obtain its general solution. The main goal is to investigate its Hyers-Ulam-Rassias (H-U-R) stability results for this equation in fuzzy $\beta-$normed spaces with appropriate two different methods of Direct and fixed point methods.

5.2 GENERAL SOLUTION

In this section, the authors derive the solution for the functional equation (5.3). Let us consider E and F as real vector spaces.

Theorem 5.1

If an odd mapping $\varphi: E \to F$ satisfies the functional equation (5.3) for all $v_1, v_2, \ldots, v_m \in E$, then the function φ is additive.

Proof: Suppose the mapping $\varphi: E \to F$ is odd, we conclude that $\varphi(-v)=-\varphi(v)$ for all $v \in E$. So that equation (5.3) becomes

$$\varphi\left(\sum_{a=1}^{m} v_a\right)+\sum_{b=1}^{m}\varphi\left(-v_b+\sum_{\substack{a=1;\\ a\neq b}}^{m} v_i\right)=(m-3)\sum_{a=1}^{m}\varphi(v_a+v_b)-\left(m^2-5m+4\right)\sum_{a=1}^{m}\varphi(v_a) \quad (5.4)$$

for all $v_1, v_2, \ldots, v_m \in E$. Taking $v_1 = v_2 = \ldots = v_m = 0$, we obtain $\varphi(0)=0$. Now, replacing $(v_1, v_2, \ldots, v_m)$ by $(v, v, 0, \ldots, 0)$ in equation (5.4), we have

$$\varphi(2v)=2\varphi(v) \quad (5.5)$$

for all $v \in E$. Replacing v by $2v$ in equation (5.5), we get

$$\varphi\left(2^2 v\right)=2^2\varphi(v) \quad (5.6)$$

for all $v \in E$. Again, replacing v by $2v$ in equation (5.6), we have

$$\varphi\left(2^3 v\right) = 2^3 \varphi(v) \tag{5.7}$$

for all $v \in E$. Therefore, we can generalize for any positive integer m such that

$$\varphi\left(2^m v\right) = 2^m \varphi(v) \tag{5.8}$$

for all $v \in E$. Replacing $(v_1, v_2, \ldots, v_m)$ by $(v_1, v_2, 0, \ldots, 0)$ in equation (5.4), we get

$$\varphi(v_1 + v_2) = \varphi(v_1) + \varphi(v_2)$$

for all $v_1, v_2 \in E$. Hence, the function φ is additive.

Theorem 5.2

If an even mapping $\varphi : E \to F$ satisfies $\varphi(0) = 0$ and (5.3) for all $v_1, v_2, \ldots, v_m \in E$, then the mapping φ is quadratic.

Proof: Suppose the mapping $\varphi : E \to F$ is even, we conclude that $\varphi(-v) = \varphi(v)$ for all $v \in E$. So that equation (5.3) becomes

$$\varphi\left(\sum_{a=1}^{m} v_a\right) + \sum_{b=1}^{m} \varphi\left(-v_b + \sum_{\substack{a=1; \\ a \neq b}}^{m} v_i\right) = (m-3)\sum_{a=1}^{m} \varphi(v_a + v_b) + \left(-m^2 + 5m - 2\right)\sum_{a=1}^{m} \varphi(v_a) \tag{5.9}$$

for all $v_1, v_2, \ldots, v_m \in E$. Now, replacing $(v_1, v_2, \ldots, v_m)$ by $(v, v, 0, \ldots, 0)$ in equation (5.9), we get

$$\varphi(2v) = 2^2 \varphi(v) \tag{5.10}$$

for all $v \in E$. Replacing v by $2v$ in equation (5.10), we have

$$\varphi\left(2^2 v\right) = 2^4 \varphi(v) \tag{5.11}$$

for all $v \in E$. Again, replacing v by $2v$ in equation (5.11), we get

$$\varphi\left(2^3 v\right) = 2^6 \varphi(v) \tag{5.12}$$

for all $v \in E$. Therefore, we can generalize for any positive integer m such that

$$\varphi(2^m v) = 2^{2m}\varphi(v) \tag{5.13}$$

for all $v \in E$. Replacing $(v_1, v_2, \ldots, v_m)$ by $(v_1, v_2, 0, \ldots, 0)$ in equation (5.13), we obtain

$$\varphi(v_1 + v_2) + \varphi(v_1 - v_2) = 2\varphi(v_1) + 2\varphi(v_2)$$

for all $v_1, v_2 \in E$. Hence, the function φ is quadratic.

Proposition 5.1

A mapping $\varphi : E \to F$ satisfies $\varphi(0) = 0$ and (5.3) for all $v_1, v_2, \ldots, v_m \in E$ if and only if there exists a mapping $Q_2 : E \times E \to F$ is symmetric bi-additive and a mapping $A_1 : E \to F$ is additive such that $\varphi(v) = Q_2(v, v) + A_1(v)$ for all $v \in E$.

Define a difference operator $\nabla\varphi : E^m \to F$ by

$$\nabla\varphi(v_1, v_2, \ldots, v_m) = \varphi\left(\sum_{a=1}^{m} v_a\right) + \sum_{b=1}^{m} \varphi\left(-v_b + \sum_{\substack{a=1; \\ a\neq b}}^{m} v_i\right) - (m-3)\sum_{a=1}^{m}\varphi(v_a + v_b)$$

$$+(m^2 - 5m + 4)\sum_{a=1}^{m}\left[\frac{\varphi(v_a) - \varphi(-v_a)}{2}\right] - (-m^2 + 5m - 2)\sum_{a=1}^{m}\left[\frac{\varphi(v_a) + \varphi(-v_a)}{2}\right]$$

for all $v_1, v_2, \ldots, v_m \in E$.

5.3 HYERS-ULAM-RASSIAS STABILITY: DIRECT METHOD

In this section, the authors investigate the H-U-R stability of the equation (5.3) in fuzzy β-normed spaces by using direct method.

Here, the authors consider that E is a real vector space, (F, N_β) is a complete fuzzy β-normed space with $0 < \beta \le 1$ and (Z, N'_β) is a fuzzy β-normed space.

5.3.1 Stability Results When φ Is Even

Theorem 5.3

Suppose an even mapping $\varphi : E \to F$ satisfies $\varphi(0) = 0$ and

$$N_\beta(\nabla\varphi(v_1, v_2, \ldots, v_m), \varepsilon) \ge N'_\beta(\tau(v_1, v_2, \ldots, v_m), \varepsilon) \quad \forall\, v_1, v_2, \ldots, v_m \in E. \tag{5.14}$$

Let a mapping $\tau : E^m \to Z$ such that

$$N'_\beta\left(\tau\left(\frac{v_1}{2^r},\frac{v_2}{2^r},\dots,\frac{v_m}{2^r}\right),\varepsilon\right) \geq N'_\beta\left(\tau(v_1,v_2,\dots,v_m),\frac{\varepsilon}{\sigma^{\beta r}}\right) \quad \forall\ v_1,v_2,\dots,v_m \in E,\ \varepsilon>0 \tag{5.15}$$

for some constant $\sigma \in \mathbb{R}$ with $0<\sigma<\frac{1}{2^2}$ and any integer $r \geq 0$. Then there is a unique quadratic mapping $Q_2 : E \to F$ satisfies

$$N_\beta(\varphi(v)-Q_2(v),\varepsilon) \geq N'_\beta\left(\tau(v,v,0,\dots,0),2^\beta\left(\sigma^{-\beta}-2^{2\beta}\right)\varepsilon\right) \tag{5.16}$$

for all $v \in E$.

Proof: Replacing $(v_1,v_2,\dots,v_m)$ by $(v,v,0,\dots,0)$ in inequality (5.14), we have

$$N_\beta(2\varphi(2v)-8\varphi(v),\varepsilon) \geq N'_\beta(\tau(v,v,0,\dots,0),\varepsilon) \tag{5.17}$$

for all $v \in E$ and $\varepsilon>0$. Replacing v by $\frac{v}{2}$ in equation (5.17), we get

$$N_\beta\left(\varphi(v)-2^2\varphi\left(\frac{v}{2}\right),\frac{\varepsilon}{2^\beta}\right) \geq N'_\beta\left(\tau\left(\frac{v}{2},\frac{v}{2},0,\dots,0\right),\varepsilon\right) \tag{5.18}$$

for all $v \in E$ and $\varepsilon>0$. Again, replacing v by $\frac{v}{2^r}$ in equation (5.18), we have

$$N_\beta\left(\varphi\left(\frac{v}{2^r}\right)-2^2\varphi\left(\frac{v}{2^{r+1}}\right),\frac{\varepsilon}{2^\beta}\right) \geq N'_\beta\left(\tau\left(\frac{v}{2^{r+1}},\frac{v}{2^{r+1}},0,\dots,0\right),\varepsilon\right) \tag{5.19}$$

for all $v \in E$ and $\varepsilon>0$. Now using (F3) and (5.19), we obtain

$$N_\beta\left(2^{2r}\varphi\left(\frac{v}{2^r}\right)-2^{2(r+1)}\varphi\left(\frac{v}{2^{r+1}}\right),\frac{2^{2r}\varepsilon}{2^\beta}\right) \geq N'_\beta\left(\tau\left(\frac{v}{2^{r+1}},\frac{v}{2^{r+1}},0,\dots,0\right),\varepsilon\right) \tag{5.20}$$

for all $v \in E$ and $\varepsilon>0$. Using inequality (5.15), we have

$$N_\beta\left(2^{2r}\varphi\left(\frac{v}{2^r}\right)-2^{2(r+1)}\varphi\left(\frac{v}{2^{r+1}}\right),\frac{2^{2r}\varepsilon}{2^\beta}\right) \geq N'_\beta\left(\tau(v,v,0,\dots,0),\frac{\varepsilon}{\sigma^{\beta(r+1)}}\right) \tag{5.21}$$

for all $v \in E$ and $\varepsilon>0$. Replacing ε by $\sigma^{\beta(r+1)}\varepsilon$ in inequality (5.21), we obtain

$$N_\beta\left(2^{2r}\varphi\left(\frac{v}{2^r}\right)-2^{2(r+1)}\varphi\left(\frac{v}{2^{r+1}}\right),\frac{\sigma^{\beta(r+1)}\,2^{2r}\varepsilon}{2^\beta}\right) \geq N'_\beta(\tau(v,v,0,\dots,0),\varepsilon) \tag{5.22}$$

for all $v \in E$ and $\varepsilon > 0$. Therefore,

$$N_\beta\left(2^{2m}\varphi\left(\frac{v}{2^m}\right)-\varphi(v),\sum_{r=0}^{m-1}\frac{\sigma^{\beta(r+1)}\,2^{2r}\varepsilon}{2^\beta}\right)=$$

$$N_\beta\left(\sum_{r=0}^{m-1}\left(2^{2(r+1)}\varphi\left(\frac{v}{2^{r+1}}\right)-2^{2r}\varphi\left(\frac{v}{2^r}\right)\right),\sum_{r=0}^{m-1}\frac{\sigma^{\beta(r+1)}\,2^{2r}\varepsilon}{2^\beta}\right)\geq$$

$$\min_{0\le r\le m-1}\left\{N_\beta\left(2^{2(r+1)}\varphi\left(\frac{v}{2^{r+1}}\right)-2^{2r}\varphi\left(\frac{v}{2^r}\right),\frac{\sigma^{\beta(r+1)}\,2^{2r}\varepsilon}{2^\beta}\right)\right\}\geq N_\beta'\left(\tau(v,v,0,\ldots,0),\varepsilon\right) \tag{5.23}$$

This implies

$$N_\beta\left(2^{2(m+s)}\varphi\left(\frac{v}{2^{(m+s)}}\right)-2^{2s}\varphi\left(\frac{v}{2^s}\right),\sum_{r=0}^{m-1}\frac{\sigma^{\beta(s+r+1)}\,2^{2\beta(s+r)}\varepsilon}{2^\beta}\right)$$

$$=N_\beta\left(\sum_{r=0}^{m-1}\left(2^{2(s+r+1)}\varphi\left(\frac{v}{2^{s+r+1}}\right)-2^{2(s+r)}\varphi\left(\frac{v}{2^{s+r}}\right)\right),\sum_{r=0}^{m-1}\frac{\sigma^{\beta(s+r+1)}\,2^{2\beta(s+r)}\varepsilon}{2^\beta}\right)$$

$$\geq \min\bigcup_{r=0}^{m-1}\left\{N_\beta\left(2^{2(s+r+1)}\varphi\left(\frac{v}{2^{(s+r+1)}}\right)-2^{2(s+r)}\varphi\left(\frac{v}{2^{(s+r)}}\right),\frac{\sigma^{\beta(s+r+1)}\,2^{2\beta(s+r)}\varepsilon}{2^\beta}\right)\right\}$$

$$\geq N_\beta'\left(\tau(v,v,0,\ldots,0),\varepsilon\right)$$

for all $v \in E$ and $\varepsilon > 0$, and $m > 0$, $s \geq 0$. Therefore,

$$N_\beta\left(2^{2(m+s)}\varphi\left(\frac{v}{2^{(m+s)}}\right)-2^{2r}\varphi\left(\frac{v}{2^r}\right),\varepsilon\right)\geq N_\beta'\left(\tau(v,v,0,\ldots,0),\frac{\varepsilon}{\sum_{r=0}^{m-1}\frac{\sigma^{\beta(s+r+1)}\,2^{2\beta(s+r)}}{2^\beta}}\right) \tag{5.24}$$

for all $v \in E$, $\varepsilon > 0$, and $m > 0$, $s \geq 0$. As the series $\sum_{r=0}^{m-1} 2^{2\beta r}\sigma^{\beta r}$ is convergent, we obtain that by considering $s = 0$ and taking $m \to \infty$ in equation (5.24) which implies $\left\{2^{2m}\varphi\left(\frac{v}{2^m}\right)\right\}$ is a Cauchy sequence in (F, N_β). So that the sequence $\left\{2^{2m}\varphi\left(\frac{v}{2^m}\right)\right\}$ converges in F.

Next, we define a mapping $Q_2 : E \to F$ by $Q_2(v) = N_\beta - \lim_{m\to\infty} 2^{2m}\varphi\left(\frac{v}{2^m}\right)$ for all $v \in E$. This shows that

$$\lim_{m\to\infty} N_\beta\left(Q_2(v) - 2^{2m}\varphi\left(\frac{v}{2^m}\right), \varepsilon\right) = 1$$

for all $v \in E$ and $\varepsilon > 0$. Replacing $(v_1, v_2, \ldots, v_m)$ by $\left(\frac{v_1}{2^i}, \frac{v_2}{2^i}, \ldots, \frac{v_m}{2^i}\right)$ in equation (5.14), we obtain

$$N_\beta\left(\nabla\varphi\left(\frac{v_1}{2^i}, \frac{v_2}{2^i}, \ldots, \frac{v_m}{2^i}\right), \varepsilon\right) \geq N_\beta^{'}\left(\tau\left(\frac{v_1}{2^i}, \frac{v_2}{2^i}, \ldots, \frac{v_m}{2^i}\right), \varepsilon\right)$$

for all $v_1, v_2, \ldots, v_m \in E$, $\varepsilon > 0$ and $i \in \mathbb{N}$. This implies

$$N_\beta\left(2^{2i}\nabla\varphi\left(\frac{v_1}{2^i}, \frac{v_2}{2^i}, \ldots, \frac{v_m}{2^i}\right), 2^{2i\beta}\varepsilon\right) \geq N_\beta^{'}\left(\tau\left(\frac{v_1}{2^i}, \frac{v_2}{2^i}, \ldots, \frac{v_m}{2^i}\right), \varepsilon\right) \tag{5.25}$$

for all $v_1, v_2, \ldots, v_m \in E$ and $\varepsilon > 0$, and $i \in \mathbb{N}$. Replacing ε by $\frac{\varepsilon}{2^{2i\beta}}$ in equation (5.25), we have

$$N_\beta\left(2^{2i}\nabla\varphi\left(\frac{v_1}{2^i}, \frac{v_2}{2^i}, \ldots, \frac{v_m}{2^i}\right), \varepsilon\right) \geq N_\beta^{'}\left(\tau\left(\frac{v_1}{2^i}, \frac{v_2}{2^i}, \ldots, \frac{v_m}{2^i}\right), \frac{\varepsilon}{2^{2i\beta}}\right)$$

for all $v_1, v_2, \ldots, v_m \in E$, $\varepsilon > 0$, and $i \in \mathbb{N}$. Using equation (5.15), we get

$$N_\beta\left(2^{2i}\nabla\varphi\left(\frac{v_1}{2^i}, \frac{v_2}{2^i}, \ldots, \frac{v_m}{2^i}\right), \varepsilon\right) \geq N_\beta^{'}\left(\tau(v_1, v_2, \ldots, v_m), \frac{\varepsilon}{(4\sigma)^{i\beta}}\right)$$

for all $v_1, v_2, \ldots, v_m \in E$, $\varepsilon > 0$, and $i \in \mathbb{N}$. Since

$$\lim_{i\to\infty} N_\beta^{'}\left(\tau(v_1, v_2, \ldots, v_m), \frac{\varepsilon}{(4\sigma)^{i\beta}}\right) = 1$$

for all $v_1, v_2, \ldots, v_m \in E$, $\varepsilon > 0$, and $i \in \mathbb{N}$. This implies that

$$N_\beta\left(\begin{array}{l}\left(Q_2\left(\sum_{a=1}^{m} v_a\right) + \sum_{b=1}^{m} Q_2\left(-v_b + \sum_{\substack{a=1;\\ a\neq b}}^{m} v_i\right)\right) - (m-3)\sum_{a=1}^{m} Q_2(v_a + v_b) + \left(m^2 - 5m + 4\right) \\ \sum_{a=1}^{m}\left[\frac{Q_2(v_a) - Q_2(-v_a)}{2}\right] - \left(-m^2 + 5m - 2\right)\sum_{a=1}^{m}\left[\frac{Q_2(v_a) + Q_2(-v_a)}{2}\right], \varepsilon\end{array}\right) = 1$$

for all $v_1, v_2, \ldots, v_m \in E$, $\varepsilon > 0$, and $i \in \mathbb{N}$. Therefore, $Q_2 : E \to F$ is a quadratic. Clearly, $\varphi(v)$ is an even, $Q_2(v)$ is an even. Taking limit $i \to \infty$ in equation (5.23), we get the result of inequality (5.16).

Next, to show that Q_2 is a unique solution, let Q_2' be another quadratic solution which satisfies the inequality (5.16), we have

$$N_\beta\left(Q_2(v)-Q_2'(v),\varepsilon\right)=N_\beta\left(2^{2n}Q_2\left(\frac{v}{2^n}\right)-2^{2n}Q_2'\left(\frac{v}{2^n}\right),\varepsilon\right)\geq$$

$$\min\left\{N_\beta\left(2^{2n}Q_2\left(\frac{v}{2^n}\right)-2^{2n}\varphi\left(\frac{v}{2^n}\right),\varepsilon\right),\ N_\beta\left(2^{2n}\varphi\left(\frac{v}{2^n}\right)-2^{2n}Q_2'\left(\frac{v}{2^n}\right),\varepsilon\right)\right\}$$

$$\geq \min\left\{N_\beta'\left(\tau\left(\frac{v}{2^i},\frac{v}{2^i},0,\ldots,0\right),\frac{\left(\sigma^{-\beta}-2^{2\beta}\right)\varepsilon}{2.2^{2i\beta}}\right),N_\beta'\left(\tau\left(\frac{v}{2^i},\frac{v}{2^i},0,\ldots,0\right),\frac{\left(\sigma^{-\beta}-2^{2\beta}\right)\varepsilon}{2.2^{2i\beta}}\right)\right\}$$

$$\geq N_\beta'\left(\tau(v,v,0,\ldots,0),\frac{\left(\sigma^{-\beta}-2^{2\beta}\right)\varepsilon}{2.(4\sigma)^{i\beta}}\right)\to 1 \text{ as } i\to\infty$$

for all $v \in E$, $\varepsilon > 0$, and all $i \in \mathbb{N}$. Hence, $Q_2 : E \to F$ is unique. Hence, the proof is completed.

Corollary 5.1

Let $w \geq 0, \gamma \in R^+$ with $\gamma > 2$. If an even mapping $\varphi : E \to F$ satisfies $\varphi(0) = 0$ and

$$N_\beta\left(\nabla\varphi(v_1,v_2,\ldots,v_m),\varepsilon\right)\geq N_\beta'\left(w\sum_{j=0}^{m}v_i^{\gamma},\varepsilon\right) \tag{5.26}$$

for all $v_1, v_2, \ldots, v_m \in E, \varepsilon > 0$, then there exists a unique quadratic mapping $Q_2 : E \to F$ satisfying

$$N_\beta\left(\varphi(v)-Q_2(v),\varepsilon\right)\geq N_\beta'\left(\tau(v,v,0,\ldots,0),2^\beta\left(2^{\gamma\beta}-2^{2\beta}\right)\varepsilon\right)$$

for all $v \in E$ and $\varepsilon > 0$.

Theorem 5.4

Suppose an even mapping $\varphi : E \to F$ satisfies $\varphi(0) = 0$ and the inequality (5.14). Let a mapping $\tau : E^m \to Z$ such that

$$N_\beta'\left(\tau(v_1,v_2,\ldots,v_m),\sigma^{\beta r}\ \varepsilon\right)\geq N_\beta'\left(\tau\left(\frac{v_1}{2^r},\frac{v_2}{2^r},\ldots,\frac{v_m}{2^r}\right),\varepsilon\right) \tag{5.27}$$

for all $v_1, v_2, \ldots, v_m \in E$, $\varepsilon > 0$ and $\varepsilon > 0$, for some constant $\sigma \in \mathbb{R}$ with $0 < \sigma < 2^2$ and any integer $r \geq 0$. Then, there is a unique quadratic mapping $Q_2 : E \to F$ that satisfies

$$N_\beta\left(\varphi(v) - Q_2(v), \varepsilon\right) \geq N'_\beta\left(\tau(v, v, 0, \ldots, 0), 2^\beta\left(2^{2\beta} - \sigma^\beta\right)\varepsilon\right) \tag{5.28}$$

for all $v \in E$ and $\varepsilon > 0$.

Proof: Replacing $(v_1, v_2, \ldots, v_m)$ by $(v, v, 0, \ldots, 0)$ in inequality (5.14), we have

$$N_\beta\left(2\varphi(2v) - 8\varphi(v), \varepsilon\right) \geq N'_\beta\left(\tau(v, v, 0, \ldots, 0), \varepsilon\right)$$

$$\Rightarrow N_\beta\left(\frac{\varphi(2v)}{2^2} - \varphi(v), \frac{\varepsilon}{2^{3\beta}}\right) \geq N'_\beta\left(\tau(v, v, 0, \ldots, 0), \varepsilon\right) \tag{5.29}$$

for all $v \in E$ and $\varepsilon > 0$. Replacing v by $2v$ in equation (5.29), we get

$$N_\beta\left(\frac{\varphi(2^2 v)}{2^4} - \frac{\varphi(2v)}{2^2}, \frac{\varepsilon}{2^{3\beta}\left(2^{2\beta}\right)}\right) \geq N'_\beta\left(\tau(2v, 2v, 0, \ldots, 0), \varepsilon\right) \tag{5.30}$$

for all $v \in E$ and $\varepsilon > 0$. For any positive integer r in $\mathbb{R}$, we have

$$N_\beta\left(\frac{\varphi(2^r v)}{2^{2r}} - \varphi(v), \frac{\varepsilon}{2^{3\beta}\left(2^{2\beta r}\right)}\right) \geq N'_\beta\left(\tau\left(2^r v, 2^r v, 0, \ldots, 0\right), \varepsilon\right) \tag{5.31}$$

for all $v \in E$ and $\varepsilon > 0$. Now, using (F3) and (5.31), we obtain

$$N_\beta\left(\frac{\varphi(2^{r+1} v)}{2^{2(r+1)}} - \frac{\varphi(2^r v)}{2^{2r}}, \frac{\varepsilon}{2^{3\beta}\left(2^{2\beta r}\right)}\right) \geq N'_\beta\left(\tau\left(2^r v, 2^r v, 0, \ldots, 0\right), \varepsilon\right) \tag{5.32}$$

for all $v \in E$, $\varepsilon > 0$. Using inequality (5.27), we have

$$N_\beta\left(\frac{\varphi(2^{r+1} v)}{2^{2(r+1)}} - \frac{\varphi(2^r v)}{2^{2r}}, \frac{\varepsilon}{2^{3\beta}\left(2^{2\beta r}\right)}\right) \geq N'_\beta\left(\tau(v, v, 0, \ldots, 0), \frac{\varepsilon}{\sigma^{\beta r}}\right) \tag{5.33}$$

for all $v \in E$, $\varepsilon > 0$. Replacing ε by $\sigma^{\beta r}\varepsilon$ in inequality (5.33), we obtain

$$N_\beta\left(\frac{\varphi(2^{r+1} v)}{2^{2(r+1)}} - \frac{\varphi(2^r v)}{2^{2r}}, \frac{\sigma^{\beta r}\varepsilon}{2^{3\beta}\left(2^{2\beta r}\right)}\right) \geq N'_\beta\left(\tau(v, v, 0, \ldots, 0), \varepsilon\right) \tag{5.34}$$

for all $v \in E$, $\varepsilon > 0$. Therefore,

$$N_\beta\left(\frac{\varphi(2^{m+1}v)}{2^{2(m+1)}} - \varphi(v), \sum_{r=0}^{m-1} \frac{\sigma^{\beta r}\varepsilon}{2^{3\beta}(2^{2\beta r})}\right) = N_\beta\left(\sum_{r=0}^{m-1}\left(\frac{\varphi(2^{r+1}v)}{2^{2(r+1)}} - \frac{\varphi(2^r v)}{2^{2r}}\right), \sum_{r=0}^{m-1} \frac{\sigma^{\beta r}\varepsilon}{2^{3\beta}(2^{2\beta r})}\right) \geq$$

$$\min_{0 \le r \le m-1}\left\{N_\beta\left(\frac{\varphi(2^{r+1}v)}{2^{2(r+1)}} - \frac{\varphi(2^r v)}{2^{2r}}, \frac{\sigma^{\beta r}\varepsilon}{2^{3\beta}(2^{2\beta r})}\right)\right\} \geq N_\beta'(\tau(v,v,0,\ldots,0),\varepsilon) \tag{5.35}$$

This implies

$$N_\beta\left(\frac{\varphi(2^{m+s}v)}{2^{2(m+s)}} - \frac{\varphi(2^s v)}{2^{2s}}, \sum_{r=0}^{m-1} \frac{\sigma^{\beta(s+r)}\varepsilon}{2^{3\beta}(2^{2\beta(s+r)})}\right)$$

$$= N_\beta\left(\sum_{r=0}^{m-1}\left(\frac{\varphi(2^{r+s}v)}{2^{2(r+s)}} - \frac{\varphi(2^s v)}{2^{2s}}\right), \sum_{r=0}^{m-1} \frac{\sigma^{\beta(s+r)}\varepsilon}{2^{3\beta}(2^{2\beta(s+r)})}\right) \geq$$

$$\min \bigcup_{r=0}^{m-1}\left\{N_\beta\left(\frac{\varphi(2^{r+s}v)}{2^{2(r+s)}} - \frac{\varphi(2^s v)}{2^{2s}}, \frac{\sigma^{\beta(s+r)}\varepsilon}{2^{3\beta}(2^{2\beta(s+r)})}\right)\right\}$$

$$\geq N_\beta'(\tau(v,v,0,\ldots,0),\varepsilon)$$

for all $v \in E$, $\varepsilon > 0$, and $m > 0$, $s \geq 0$. Therefore,

$$N_\beta\left(\frac{\varphi(2^{m+s}v)}{2^{2(m+s)}} - \frac{\varphi(2^s v)}{2^{2s}}, \varepsilon\right) \geq N_\beta'\left(\tau(v,v,0,\ldots,0), \frac{\varepsilon}{\sum_{r=0}^{m-1} \frac{\sigma^{\beta(s+r)}}{2^{3\beta}(2^{2\beta(s+r)})}}\right) \tag{5.36}$$

for all $v \in E$, $\varepsilon > 0$, and $m > 0$, $s \geq 0$. As the series $\sum_{r=0}^{m-1} \frac{\sigma^{\beta r}}{2^{2\beta r}}$ is convergent, we obtain that by considering $s = 0$ and $m \to \infty$ in equation (5.36) which yields that the sequence $\left\{\frac{\varphi(2^m v)}{2^{2m}}\right\}$ is a Cauchy in (F, N_β). Therefore, the sequence $\left\{\frac{\varphi(2^m v)}{2^{2m}}\right\}$ converges in F.

Next, we define a mapping $Q_2 : E \to F$ by $Q_2(v) = N_\beta - \lim_{m\to\infty} \frac{\varphi(2^m v)}{2^{2m}}$ for all $v \in E$. This shows that

$$\lim_{m\to\infty} N_\beta\left(Q_2(v) - \frac{\varphi(2^m v)}{2^{2m}}, \varepsilon\right) = 1$$

for all $v \in E$ and $\varepsilon > 0$. Replacing $(v_1, v_2, \ldots, v_m)$ by $(2^i v_1, 2^i v_2, \ldots 2^i v_m)$ in equation (5.14), we obtain

$$N_\beta\left(\nabla\varphi\left(2^i v_1, 2^i v_2, \ldots 2^i v_m\right), \varepsilon\right) \geq N_\beta'\left(\tau\left(2^i v_1, 2^i v_2, \ldots 2^i v_m\right), \varepsilon\right)$$

for all $\forall\ v_1, v_2, \ldots, v_m \in E$ and $\varepsilon > 0$ and $i \in \mathbb{N}$. This implies

$$N_\beta\left(\frac{1}{2^{2i}}\nabla\varphi\left(2^i v_1, 2^i v_2, \ldots 2^i v_m\right), 2^{2i\beta}\varepsilon\right) \geq N_\beta'\left(\tau\left(2^i v_1, 2^i v_2, \ldots 2^i v_m\right), \varepsilon\right) \quad (5.37)$$

for all $v_1, v_2, \ldots, v_m \in E$, $\varepsilon > 0$, and $i \in \mathbb{N}$. Replacing ε by $\frac{\varepsilon}{2^{2i\beta}}$ in equation (5.37), we have

$$N_\beta\left(\frac{1}{2^{2i}}\nabla\varphi\left(2^i v_1, 2^i v_2, \ldots 2^i v_m\right), \varepsilon\right) \geq N_\beta'\left(\tau\left(2^i v_1, 2^i v_2, \ldots 2^i v_m\right), \frac{\varepsilon}{2^{2i\beta}}\right)$$

for all $v_1, v_2, \ldots, v_m \in E$, $\varepsilon > 0$, and $i \in \mathbb{N}$. Using (5.27), we get

$$N_\beta\left(\frac{1}{2^{2i}}\nabla\varphi\left(2^i v_1, 2^i v_2, \ldots 2^i v_m\right), \varepsilon\right) \geq N_\beta'\left(\tau\left(v_1, v_2, \ldots, v_m\right), \frac{\varepsilon}{(4\sigma)^{i\beta}}\right)$$

for all $v_1, v_2, \ldots, v_m \in E$, $\varepsilon > 0$, and $i \in \mathbb{N}$. Since

$$\lim_{i\to\infty} N_\beta'\left(\tau\left(v_1, v_2, \ldots, v_m\right), \frac{\varepsilon}{(4\sigma)^{i\beta}}\right) = 1$$

for all $v_1, v_2, \ldots, v_m \in E$, $\varepsilon > 0$, and $i \in \mathbb{N}$. This implies that

$$N_\beta\left(\begin{array}{c} Q_2\left(\sum_{a=1}^{m} v_a\right) + \sum_{b=1}^{m} Q_2 Q_2\left(-v_b + \sum_{\substack{a=1;\\ a\neq b}}^{m} v_i\right) - (m-3)\sum_{a=1}^{m} Q_2\left(v_a + v_b\right) + \left(m^2 - 5m + 4\right) \\ \sum_{a=1}^{m}\left[\frac{Q_2(v_a) - Q_2(-v_a)}{2}\right] - \left(-m^2 + 5m - 2\right)\sum_{a=1}^{m}\left[\frac{Q_2(v_a) + Q_2(-v_a)}{2}\right], \varepsilon \end{array}\right) = 1$$

for all $v_1, v_2, \ldots, v_m \in E$, $\varepsilon > 0$, and $i \in \mathbb{N}$. Therefore, $Q_2 : E \to F$ is a quadratic. Since $\varphi(v)$ is an even, $Q_2(v)$ is an even. Taking limit $i \to \infty$ in equation (5.35), we get the result of inequality (5.28).

Now, to show that the function Q_2 is unique, let Q_2' be another quadratic function which satisfies the inequality (5.28), we have

$$N_\beta\left(Q_2(v)-Q_2^{'}(v),\varepsilon\right)=N_\beta\left(\frac{Q_2(2^n v)}{2^{2n}}-\frac{Q_2^{'}(2^n v)}{2^{2n}},\varepsilon\right)\geq$$

$$\min\left\{N_\beta\left(\frac{Q_2(2^n v)}{2^{2n}}-\frac{\varphi(2^n v)}{2^{2n}},\varepsilon\right),\ N_\beta\left(\frac{\varphi(2^n v)}{2^{2n}}-\frac{Q_2^{'}(2^n v)}{2^{2n}},\varepsilon\right)\right\}\geq$$

$$\min\left\{N_\beta^{'}\left(\tau(2^i v,2^i v,0,\ldots,0),\frac{\left(2^{2\beta}-\sigma^\beta\right)\varepsilon}{2.2^{2i\beta}}\right),N_\beta^{'}\left(\tau(2^i v,2^i v,0,\ldots,0),\frac{\left(2^{2\beta}-\sigma^\beta\right)\varepsilon}{2.2^{2i\beta}}\right)\right\}\geq$$

$$N_\beta^{'}\left(\tau(v,v,0,\ldots,0),\frac{\left(2^{2\beta}-\sigma^\beta\right)\varepsilon}{2.(4\sigma)^{i\beta}}\right)\rightarrow 1\ as\ \rightarrow\infty$$

for all $v\in E$, $\varepsilon>0$, and all $i\in\mathbb{N}$. Hence, $Q_2:E\rightarrow F$ is unique. Hence, the proof is completed.

Corollary 5.2

Let $w\geq 0,\gamma\in R^+$ with $0<\gamma<2$. If an even mapping $\varphi:E\rightarrow F$ satisfies $\varphi(0)=0$ and (5.26) for all $v_1,v_2,\ldots,v_m\in E,\varepsilon>0$, then there exists a unique quadratic mapping $Q_2:E\rightarrow F$ that satisfies

$$N_\beta(\varphi(v)-Q_2(v),\varepsilon)\geq N_\beta^{'}\left(\tau(v,v,0,\ldots,0),2^\beta\left(2^{2\beta}-2^{\gamma\beta}\right)\varepsilon\right)$$

for all $v\in E$ and $\varepsilon>0$.

5.3.2 Stability Results When φ Is Odd

Theorem 5.5

Suppose an odd mapping $\varphi:E\rightarrow F$ satisfies (5.14) for all $v_1,v_2,\ldots,v_m\in E,\varepsilon>0$. Let a mapping $\tau:E^m\rightarrow Z$ such that the inequality (5.15) holds for all $v_1,v_2,\ldots,v_m\in E$ and $\varepsilon>0$, and for some constant $\sigma\in\mathbb{R}$ with $0<\sigma<\frac{1}{2}$ and any integer $r\geq 0$, then there exists a unique additive mapping $A_1:E\rightarrow F$ that satisfies

$$N_\beta(\varphi(v)-A_1(v),\varepsilon)\geq N_\beta^{'}\left(\tau(v,v,0,\ldots,0),2^\beta\left(\sigma^{-\beta}-2^\beta\right)\varepsilon\right) \tag{5.38}$$

for all $v\in E$ and $\varepsilon>0$.

Proof: Replacing $(v_1, v_2, \ldots, v_m)$ by $(v, v, 0, \ldots, 0)$ in inequality (5.14), we have

$$N_\beta\left(2\varphi(2v) - 4\varphi(v), \varepsilon\right) \geq N_\beta'\left(\tau(v, v, 0, \ldots, 0), \varepsilon\right) \tag{5.39}$$

for all $v \in E$ and $\varepsilon > 0$. Replacing v by $\frac{v}{2}$ in equation (5.39), we get

$$N_\beta\left(\varphi(v) - 2\varphi\left(\frac{v}{2}\right), \frac{\varepsilon}{2^\beta}\right) \geq N_\beta'\left(\tau\left(\frac{v}{2}, \frac{v}{2}, 0, \ldots, 0\right), \varepsilon\right) \tag{5.40}$$

for all $v \in E$ and $\varepsilon > 0$. Again, replacing v by $\frac{v}{2^r}$ in equation (5.40), we have

$$N_\beta\left(\varphi\left(\frac{v}{2^r}\right) - 2\varphi\left(\frac{v}{2^{r+1}}\right), \frac{\varepsilon}{2^\beta}\right) \geq N_\beta'\left(\tau\left(\frac{v}{2^{r+1}}, \frac{v}{2^{r+1}}, 0, \ldots, 0\right), \varepsilon\right) \tag{5.41}$$

for all $v \in E$ and $\varepsilon > 0$. Now, using (F3) and (5.41), we obtain

$$N_\beta\left(2^r\varphi\left(\frac{v}{2^r}\right) - 2^{(r+1)}\varphi\left(\frac{v}{2^{r+1}}\right), \frac{2^r\varepsilon}{2^\beta}\right) \geq N_\beta'\left(\tau\left(\frac{v}{2^{r+1}}, \frac{v}{2^{r+1}}, 0, \ldots, 0\right), \varepsilon\right) \tag{5.42}$$

for all $v \in E$, $\varepsilon > 0$. Using inequality (5.15), we have

$$N_\beta\left(2^r\varphi\left(\frac{v}{2^r}\right) - 2^{(r+1)}\varphi\left(\frac{v}{2^{r+1}}\right), \frac{2^r\varepsilon}{2^\beta}\right) \geq N_\beta'\left(\tau(v, v, 0, \ldots, 0), \frac{\varepsilon}{\sigma^{\beta(r+1)}}\right) \tag{5.43}$$

for all $v \in E$, $\varepsilon > 0$. Replacing ε by $\sigma^{\beta(r+1)}\varepsilon$ in inequality (5.43), we attain

$$N_\beta\left(2^r\varphi\left(\frac{v}{2^r}\right) - 2^{(r+1)}\varphi\left(\frac{v}{2^{r+1}}\right), \frac{\sigma^{\beta(r+1)}\, 2^r\varepsilon}{2^\beta}\right) \geq N_\beta'\left(\tau(v, v, 0, \ldots, 0), \varepsilon\right) \tag{5.44}$$

for all $v \in E$, $\varepsilon > 0$. Therefore,

$$\begin{aligned}
&N_\beta\left(2^m\varphi\left(\frac{v}{2^m}\right) - \varphi(v), \sum_{r=0}^{m-1}\frac{\sigma^{\beta(r+1)}\, 2^r\varepsilon}{2^\beta}\right) = \\
&N_\beta\left(\sum_{r=0}^{m-1}\left(2^{(r+1)}\varphi\left(\frac{v}{2^{r+1}}\right) - 2^r\varphi\left(\frac{v}{2^r}\right)\right), \sum_{r=0}^{m-1}\frac{\sigma^{\beta(r+1)}\, 2^r\varepsilon}{2^\beta}\right) \geq \\
&\min_{0\leq r\leq m-1}\left\{N_\beta\left(2^{(r+1)}\varphi\left(\frac{v}{2^{r+1}}\right) - 2^r\varphi\left(\frac{v}{2^r}\right), \frac{\sigma^{\beta(r+1)}\, 2^r\varepsilon}{2^\beta}\right)\right\} \geq \\
&N_\beta'\left(\tau(v, v, 0, \ldots, 0), \varepsilon\right)
\end{aligned} \tag{5.45}$$

This implies

$$N_\beta\left(2^{(m+s)}\varphi\left(\frac{v}{2^{(m+s)}}\right)-2^s\varphi\left(\frac{v}{2^s}\right),\sum_{r=0}^{m-1}\frac{\sigma^{\beta(s+r+1)}\,2^{\beta(s+r)}\varepsilon}{2^\beta}\right)=$$

$$N_\beta\left(\sum_{r=0}^{m-1}\left(2^{(s+r+1)}\varphi\left(\frac{v}{2^{s+r+1}}\right)-2^{(s+r)}\varphi\left(\frac{v}{2^{s+r}}\right)\right),\sum_{r=0}^{m-1}\frac{\sigma^{\beta(s+r+1)}\,2^{\beta(s+r)}\varepsilon}{2^\beta}\right)$$

$$\geq \min\bigcup_{r-0}^{m-1}\left\{N_\beta\left(2^{(s+r+1)}\varphi\left(\frac{v}{2^{(s+r+1)}}\right)-2^{(s+r)}\varphi\left(\frac{v}{2^{(s+r)}}\right),\frac{\sigma^{\beta(s+r+1)}\,2^{\beta(s+r)}\varepsilon}{2^\beta}\right)\right\}$$

$$\geq N_\beta'\left(\tau(v,v,0,\ldots,0),\varepsilon\right)$$

for all $v\in E$, $\varepsilon>0$, and $m>0$, $s\geq 0$. Therefore,

$$N_\beta\left(2^{(m+s)}\varphi\left(\frac{v}{2^{(m+s)}}\right)-2^r\varphi\left(\frac{v}{2^r}\right),\varepsilon\right)\geq N_\beta'\left(\tau(v,v,0,\ldots,0),\frac{\varepsilon}{\sum_{r=0}^{m-1}\frac{\sigma^{\beta(s+r+1)}\,2^{\beta(s+r)}}{2^\beta}}\right) \tag{5.46}$$

for all $v\in E$, $\varepsilon>0$, and $m>0$, $s\geq 0$. As the series $\sum_{r=0}^{m-1}2^{\beta r}\sigma^{\beta r}$ is convergent, we obtain that by considering $s=0$ and taking $m\to\infty$ in equation (5.46) which yields that the sequence $\left\{2^m\varphi\left(\frac{v}{2^m}\right)\right\}$ is a Cauchy in (F,N_β). Therefore, the sequence $\left\{\frac{\varphi(2^m v)}{2^{2m}}\right\}$ converges in F.

Next, we define a mapping $A_1:E\to F$ by $A_1(v)=N_\beta-\lim_{m\to\infty}2^m\varphi\left(\frac{v}{2^m}\right)$ for all $v\in E$. This shows that

$$\lim_{m\to\infty}N_\beta\left(A_1(v)-2^m\varphi\left(\frac{v}{2^m}\right),\varepsilon\right)=1$$

for all $v\in E$ and $\varepsilon>0$. Replacing $(v_1,v_2,\ldots,v_m)$ by $\left(\frac{v_1}{2^i},\frac{v_2}{2^i},\ldots,\frac{v_m}{2^i}\right)$ in equation (5.14), we obtain

$$N_\beta\left(\nabla\varphi\left(\frac{v_1}{2^i},\frac{v_2}{2^i},\ldots,\frac{v_m}{2^i}\right),\varepsilon\right)\geq N_\beta'\left(\tau\left(\frac{v_1}{2^i},\frac{v_2}{2^i},\ldots,\frac{v_m}{2^i}\right),\varepsilon\right)\quad\forall\ v_1,v_2,\ldots,v_m\in E,\ \varepsilon>0,$$

$i\in\mathbb{N}$.

This implies

$$N_\beta\left(2^i\nabla\varphi\left(\frac{v_1}{2^i},\frac{v_2}{2^i},\ldots,\frac{v_m}{2^i}\right),2^{i\beta}\varepsilon\right)\geq N_\beta'\left(\tau\left(\frac{v_1}{2^i},\frac{v_2}{2^i},\ldots,\frac{v_m}{2^i}\right),\varepsilon\right) \tag{5.47}$$

for all $v_1,v_2,\ldots,v_m\in E$, $\varepsilon>0$, and $i\in\mathbb{N}$. Replacing ε by $\frac{\varepsilon}{2^{i\beta}}$ in equation (5.47), we have

$$N_\beta\left(2^i\nabla\varphi\left(\frac{v_1}{2^i},\frac{v_2}{2^i},\ldots,\frac{v_m}{2^i}\right),\varepsilon\right)\geq N_\beta'\left(\tau\left(\frac{v_1}{2^i},\frac{v_2}{2^i},\ldots,\frac{v_m}{2^i}\right),\frac{\varepsilon}{2^{i\beta}}\right)$$

for all $v_1,v_2,\ldots,v_m\in E$, $\varepsilon>0$, and $i\in\mathbb{N}$. Using (5.15), we get

$$N_\beta\left(2^i\nabla\varphi\left(\frac{v_1}{2^i},\frac{v_2}{2^i},\ldots,\frac{v_m}{2^i}\right),\varepsilon\right)\geq N_\beta'\left(\tau(v_1,v_2,\ldots,v_m),\frac{\varepsilon}{(2\sigma)^{i\beta}}\right)$$

for all $v_1,v_2,\ldots,v_m\in E$, $\varepsilon>0$, and $i\in\mathbb{N}$. Since

$$\lim_{i\to\infty}N_\beta'\left(\tau(v_1,v_2,\ldots,v_m),\frac{\varepsilon}{(2\sigma)^{i\beta}}\right)=1$$

for all $v_1,v_2,\ldots,v_m\in E$, $\varepsilon>0$, and $i\in\mathbb{N}$. This implies that

$$N_\beta\left(\begin{array}{l}A_1\left(\sum_{a=1}^{m}v_a\right)+\sum_{b=1}^{m}A_1\left(-v_b+\sum_{\substack{a=1;\\a\neq b}}^{m}v_i\right)-\\(m-3)\sum_{a=1}^{m}A_1(v_a+v_b)+\left(m^2-5m+4\right)\sum_{a=1}^{m}\left[\frac{A_1(v_a)-A_1(-v_a)}{2}\right]-\\\left(-m^2+5m-2\right)\sum_{a=1}^{m}\left[\frac{A_1(v_a)+A_1(-v_a)}{2}\right],\varepsilon\end{array}\right)=1$$

for all $v_1,v_2,\ldots,v_m\in E$, $\varepsilon>0$, and $i\in\mathbb{N}$. Therefore, $A_1:E\to F$ is an additive mapping. Since $\varphi(v)$ is an odd mapping, $A_1(v)$ is an odd mapping. Taking limit $i\to\infty$ in equation (5.45), we get the result of inequality (5.38).

Now, to show that the function A_1 is unique, let A_1' be another additive function which satisfying (5.38), we obtain that

$$N_\beta\left(A_1(v)-A_1^{'}(v),\varepsilon\right)=N_\beta\left(2^n A_1\left(\frac{v}{2^n}\right)-2^n A_1^{'}\left(\frac{v}{2^n}\right),\varepsilon\right)\geq$$

$$\min\left\{N_\beta\left(2^n A_1\left(\frac{v}{2^n}\right)-2^n\varphi\left(\frac{v}{2^n}\right),\varepsilon\right),\ N_\beta\left(2^n\varphi\left(\frac{v}{2^n}\right)-2^n A_1^{'}\left(\frac{v}{2^n}\right),\varepsilon\right)\right\}\geq$$

$$\min\left\{N_\beta^{'}\left(\tau\left(\frac{v}{2^i},\frac{v}{2^i},0,\ldots,0\right),\frac{\left(\sigma^{-\beta}-2^\beta\right)\varepsilon}{2.2^{i\beta}}\right),N_\beta^{'}\left(\tau\left(\frac{v}{2^i},\frac{v}{2^i},0,\ldots,0\right),\frac{\left(\sigma^{-\beta}-2^\beta\right)\varepsilon}{2.2^{i\beta}}\right)\right\}$$

$$\geq N_\beta^{'}\left(\tau(v,v,0,\ldots,0),\frac{\left(\sigma^{-\beta}-2^\beta\right)\varepsilon}{2.(2\sigma)^{i\beta}}\right)\to 1 \text{ as } i\to\infty$$

for all $v\in E$, $\varepsilon>0$, and all $i\in\mathbb{N}$. Hence, $A_1:E\to F$ is unique. Hence the proof is completed.

Corollary 5.3

Let $w\geq 0,\gamma\in R^+$ with $\gamma>1$. If an odd mapping $\varphi:E\to F$ which satisfies the inequality (5.26) holds for every $v_1,v_2,\ldots,v_m\in E,\varepsilon>0$, then there exists a unique additive mapping $A_1:E\to F$ that satisfies

$$N_\beta(\varphi(v)-A_1(v),\varepsilon)\geq N_\beta^{'}\left(\tau(v,v,0,\ldots,0),2^\beta\left(2^{\gamma\beta}-2^\beta\right)\varepsilon\right)$$

for all $v\in E$ and $\varepsilon>0$.

Theorem 5.6

Suppose an odd mapping $\varphi:E\to F$ satisfies (5.14). Let a mapping $\tau:E^m\to Z$ such that the inequality (5.27) holds for all $v_1,v_2,\ldots,v_m\in E$, $\varepsilon>0$, and for some constant $\sigma\in\mathbb{R}$ with $0<\sigma<2$ and any integer $r\geq 0$. Then there exists a unique additive function $A_1:E\to F$ that satisfies

$$N_\beta(\varphi(v)-A_1(v),\varepsilon)\geq N_\beta^{'}\left(\tau(v,v,0,\ldots,0),2^\beta\left(2^\beta-\sigma^\beta\right)\varepsilon\right)\tag{5.48}$$

for all $v\in E$ and $\varepsilon>0$.

Proof: Replacing $(v_1,v_2,\ldots,v_m)$ by $(v,v,0,\ldots,0)$ in inequality (5.14), we have

$$N_\beta(2\varphi(2v)-4\varphi(v),\varepsilon)\geq N_\beta^{'}(\tau(v,v,0,\ldots,0),\varepsilon)$$

$$\Rightarrow N_\beta\left(\frac{\varphi(2v)}{2}-\varphi(v),\ \frac{\varepsilon}{2^{2\beta}}\right)\geq N_\beta^{'}(\tau(v,v,0,\ldots,0),\varepsilon)\tag{5.49}$$

for all $v \in E$ and $\varepsilon > 0$. Replacing v by $2v$ in equation (5.49), we get

$$N_\beta\left(\frac{\varphi(2^2 v)}{2^2} - \frac{\varphi(2v)}{2}, \frac{\varepsilon}{2^{2\beta}(2^\beta)}\right) \geq N'_\beta(\tau(2v, 2v, 0, \ldots, 0), \varepsilon) \quad (5.50)$$

for all $v \in E$ and $\varepsilon > 0$. For any positive integer r in $\mathbb{R}$, we have

$$N_\beta\left(\frac{\varphi(2^r v)}{2^2} - \varphi(v), \frac{\varepsilon}{2^{2\beta}(2^{\beta r})}\right) \geq N'_\beta(\tau(2^r v, 2^r v, 0, \ldots, 0), \varepsilon) \quad (5.51)$$

for all $v \in E$ and $\varepsilon > 0$. Now, using (F3) and (5.51), we obtain

$$N_\beta\left(\frac{\varphi(2^{r+1} v)}{2^{(r+1)}} - \frac{\varphi(2^r v)}{2^r}, \frac{\varepsilon}{2^{2\beta}(2^{\beta r})}\right) \geq N'_\beta(\tau(2^r v, 2^r v, 0, \ldots, 0), \varepsilon) \quad (5.52)$$

for all $v \in E$, $\varepsilon > 0$. Using inequality (5.27), we have

$$N_\beta\left(\frac{\varphi(2^{r+1} v)}{2^{(r+1)}} - \frac{\varphi(2^r v)}{2^r}, \frac{\varepsilon}{2^{2\beta}(2^{\beta r})}\right) \geq N'_\beta\left(\tau(v, v, 0, \ldots, 0), \frac{\varepsilon}{\sigma^{\beta r}}\right) \quad (5.53)$$

for all $v \in E$, $\varepsilon > 0$. Replacing ε by $\sigma^{\beta r}\varepsilon$ in inequality (5.53), we obtain

$$N_\beta\left(\frac{\varphi(2^{r+1} v)}{2^{(r+1)}} - \frac{\varphi(2^r v)}{2^r}, \frac{\sigma^{\beta r}\varepsilon}{2^{2\beta}(2^{\beta r})}\right) \geq N'_\beta(\tau(v, v, 0, \ldots, 0), \varepsilon) \quad (5.54)$$

for all $v \in E$, $\varepsilon > 0$. Therefore,

$$N_\beta\left(\frac{\varphi(2^{m+1} v)}{2^{(m+1)}} - \varphi(v), \sum_{r=0}^{m-1} \frac{\sigma^{\beta r}\varepsilon}{2^{2\beta}(2^{\beta r})}\right) = N_\beta\left(\sum_{r=0}^{m-1}\left(\frac{\varphi(2^{r+1} v)}{2^{(r+1)}} - \frac{\varphi(2^r v)}{2^r}\right), \sum_{r=0}^{m-1} \frac{\sigma^{\beta r}\varepsilon}{2^{2\beta}(2^{\beta r})}\right)$$

$$\geq \min_{0 \leq r \leq m-1}\left\{N_\beta\left(\frac{\varphi(2^{r+1} v)}{2^{(r+1)}} - \frac{\varphi(2^r v)}{2^r}, \frac{\sigma^{\beta r}\varepsilon}{2^{2\beta}(2^{\beta r})}\right)\right\} \geq N'_\beta(\tau(v, v, 0, \ldots, 0), \varepsilon) \quad (5.55)$$

This implies

$$N_\beta\left(\frac{\varphi(2^{m+s}v)}{2^{(m+s)}}-\frac{\varphi(2^s v)}{2^s},\sum_{r=0}^{m-1}\frac{\sigma^{\beta(s+r)}\varepsilon}{2^{2\beta}\left(2^{\beta(s+r)}\right)}\right)=$$

$$N_\beta\left(\sum_{r=0}^{m-1}\left(\frac{\varphi(2^{r+s}v)}{2^{(r+s)}}-\frac{\varphi(2^s v)}{2^s}\right),\sum_{r=0}^{m-1}\frac{\sigma^{\beta(s+r)}\varepsilon}{2^{2\beta}\left(2^{\beta(s+r)}\right)}\right)\geq$$

$$\min\bigcup_{r=0}^{m-1}\left\{N_\beta\left(\frac{\varphi(2^{r+s}v)}{2^{(r+s)}}-\frac{\varphi(2^s v)}{2^s},\frac{\sigma^{\beta(s+r)}\varepsilon}{2^{2\beta}\left(2^{\beta(s+r)}\right)}\right)\right\}\geq$$

$$N'_\beta\left(\tau(v,v,0,\ldots,0),\varepsilon\right)$$

for all $v\in E$, $\varepsilon>0$, and $m>0$, $s\geq 0$. Therefore,

$$N_\beta\left(\frac{\varphi(2^{m+s}v)}{2^{(m+s)}}-\frac{\varphi(2^s v)}{2^s},\varepsilon\right)\geq N'_\beta\left(\tau(v,v,0,\ldots,0),\frac{\varepsilon}{\sum_{r=0}^{m-1}\frac{\sigma^{\beta(s+r)}}{2^{2\beta}\left(2^{\beta(s+r)}\right)}}\right)\tag{5.56}$$

for all $v\in E$, $\varepsilon>0$, and $m>0$, $s\geq 0$. As the series $\sum_{r=0}^{m-1}\frac{\sigma^{\beta r}}{2^{\beta r}}$ is convergent, we obtain that by considering $s=0$ and $m\to\infty$ in equation (5.56) which yields that the sequence $\left\{\frac{\varphi(2^m v)}{2^m}\right\}$ is a Cauchy in (F,N_β). Therefore, the sequence $\left\{\frac{\varphi(2^m v)}{2^m}\right\}$ converges in F.

Next, we define a mapping $A_1:E\to F$ by $A_1(v)=N_\beta-\lim_{m\to\infty}\frac{\varphi(2^m v)}{2^m}$ for all $v\in E$. This shows that

$$\lim_{m\to\infty}N_\beta\left(A_1(v)-\frac{\varphi(2^m v)}{2^m},\varepsilon\right)=1$$

for all $v\in E$ and $\varepsilon>0$. Replacing $(v_1,v_2,\ldots,v_m)$ by $\left(2^i v_1,2^i v_2,\ldots 2^i v_m\right)$ in equation (5.14), we obtain

$$N_\beta\left(\nabla\varphi\left(2^i v_1,2^i v_2,\ldots 2^i v_m\right),\varepsilon\right)\geq N'_\beta\left(\tau\left(2^i v_1,2^i v_2,\ldots 2^i v_m\right),\varepsilon\right)$$

for all $\forall$ $v_1,v_2,\ldots,v_m\in E$, $\varepsilon>0$ and $i\in\mathbb{N}$. This implies

$$N_\beta\left(\frac{1}{2^i}\nabla\varphi\left(2^i v_1, 2^i v_2, \ldots 2^i v_m\right), 2^{i\beta}\varepsilon\right) \geq N'_\beta\left(\tau\left(2^i v_1, 2^i v_2, \ldots 2^i v_m\right), \varepsilon\right) \quad (5.57)$$

for all $v_1, v_2, \ldots, v_m \in E$, $\varepsilon > 0$, and $i \in \mathbb{N}$. Replacing ε by $\frac{\varepsilon}{2^{i\beta}}$ in equation (5.57), we have

$$N_\beta\left(\frac{1}{2^i}\nabla\varphi\left(2^i v_1, 2^i v_2, \ldots 2^i v_m\right), \varepsilon\right) \geq N'_\beta\left(\tau\left(2^i v_1, 2^i v_2, \ldots 2^i v_m\right), \frac{\varepsilon}{2^{i\beta}}\right)$$

for all $v_1, v_2, \ldots, v_m \in E$, $\varepsilon > 0$, and $i \in \mathbb{N}$. Using (5.27), we get

$$N_\beta\left(\frac{1}{2^i}\nabla\varphi\left(2^i v_1, 2^i v_2, \ldots 2^i v_m\right), \varepsilon\right) \geq N'_\beta\left(\tau\left(v_1, v_2, \ldots, v_m\right), \frac{\varepsilon}{(2\sigma)^{i\beta}}\right)$$

for all $v_1, v_2, \ldots, v_m \in E$, $\varepsilon > 0$, and $i \in \mathbb{N}$. Since

$$\lim_{i\to\infty} N'_\beta\left(\tau\left(v_1, v_2, \ldots, v_m\right), \frac{\varepsilon}{(2\sigma)^{i\beta}}\right) = 1$$

for all $v_1, v_2, \ldots, v_m \in E$, $\varepsilon > 0$, and $i \in \mathbb{N}$. This implies that

$$N_\beta\left(\begin{array}{l} A_1\left(\sum_{a=1}^{m} v_a\right) + \sum_{b=1}^{m} A_1\left(-v_b + \sum_{\substack{a=1;\\ a\neq b}}^{m} v_i\right) - \\ (m-3)\sum_{a=1}^{m} A_1\left(v_a + v_b\right) + \left(m^2 - 5m + 4\right)\sum_{a=1}^{m}\left[\frac{A_1\left(v_a\right) - A_1\left(-v_a\right)}{2}\right] - \\ \left(-m^2 + 5m - 2\right)\sum_{a=1}^{m}\left[\frac{A_1\left(v_a\right) + A_1\left(-v_a\right)}{2}\right], \varepsilon \end{array}\right) = 1$$

for all $v_1, v_2, \ldots, v_m \in E$, $\varepsilon > 0$, and $i \in \mathbb{N}$. Therefore, $A_1 : E \to F$ is an additive. As the function $\varphi(v)$ is odd, $A_1(v)$ is an odd function. Taking limit $i \to \infty$ in equation (5.55), we get the result of inequality (5.48).

Now, to show that the function A_1 is unique, let A_1' be another additive function which satisfies (5.48), we obtain that

$$N_\beta\left(A_1(v)-A_1'(v),\varepsilon\right)=N_\beta\left(\frac{A_1(2^n v)}{2^n}-\frac{A_1'(2^n v)}{2^n},\varepsilon\right)\geq$$

$$\min\left\{N_\beta\left(\frac{A_1(2^n v)}{2^n}-\frac{\varphi(2^n v)}{2^n},\varepsilon\right),\ N_\beta\left(\frac{\varphi(2^n v)}{2^n}-\frac{A_1'(2^n v)}{2^n},\varepsilon\right)\right\}\geq$$

$$\min\left\{N_\beta'\left(\tau(2^i v,2^i v,0,\ldots,0),\frac{(2^\beta-\sigma^\beta)\varepsilon}{2.2^{i\beta}}\right),N_\beta'\left(\tau(2^i v,2^i v,0,\ldots,0),\frac{(2^\beta-\sigma^\beta)\varepsilon}{2.2^{i\beta}}\right)\right\}$$

$$\geq N_\beta'\left(\tau(v,v,0,\ldots,0),\frac{(2^\beta-\sigma^\beta)\varepsilon}{2.(2\sigma)^{i\beta}}\right)\rightarrow \text{as } i\rightarrow\infty$$

for all $v\in E$, $\varepsilon>0$, and all $i\in\mathbb{N}$. Hence, $A_1:E\rightarrow F$ is unique. Hence, the proof is completed.

Corollary 5.4

Let $w\geq 0,\gamma\in R^+$ with $0<\gamma<1$. If an odd mapping $\varphi:E\rightarrow F$ satisfies (5.26) for every $v_1,v_2,\ldots,v_m\in E,\varepsilon>0$, then there exists a unique additive mapping $A_1:E\rightarrow F$ that satisfies

$$N_\beta(\varphi(v)-A_1(v),\varepsilon)\geq N_\beta'\left(\tau(v,v,0,\ldots,0),2^\beta\left(2^\beta-2^{\gamma\beta}\right)\varepsilon\right)$$

for all $v\in E$ and $\varepsilon>0$.

5.4 HYERS-ULAM-RASSIAS STABILITY: FIXED-POINT METHOD

In this section, the authors examined the H-U-R stability of the equation (5.3) in fuzzy β-normed spaces by fixed point method.

5.4.1 Stability Results When φ Is Even

Theorem 5.7

Let a function $\tau:E^m\rightarrow[0,\infty)$ and there is a Lipschitz constant L, $0<L<1$, which satisfies

$$\tau\left(2^i v_1,2^i v_2,\ldots,2^i v_m\right)\leq 2^{2i\beta}L\tau(v_1,v_2,\ldots,v_m)\quad \forall\ v_1,v_2,\ldots,v_m\in E. \tag{5.58}$$

for all $v_1, v_2, \ldots, v_m \in E$ and $\varepsilon > 0$. Suppose an even function $\varphi : E \to F$ satisfying $\varphi(0) = 0$ and

$$N_\beta\left(\nabla\varphi(v_1, v_2, \ldots, v_m), \varepsilon\right) \geq \frac{\varepsilon}{\varepsilon + \tau(v_1, v_2, \ldots, v_m)} \tag{5.59}$$

for all $v_1, v_2, \ldots, v_m \in E$ and $\varepsilon > 0$, then there is a unique quadratic mapping $Q_2 : E \to F$ that satisfies

$$N_\beta\left(\varphi(v) - Q_2(v), \varepsilon\right) \geq \frac{2^{3\beta}(1-L)\varepsilon}{2^{3\beta}(1-L)\varepsilon + \tau(v, v, 0, \ldots, 0)} \tag{5.60}$$

for all $v \in E$ and $\varepsilon > 0$.

Proof: Replacing $(v_1, v_2, \ldots, v_m)$ by $(v, v, 0, \ldots, 0)$ in equation (5.59), we have

$$N_\beta\left(2\varphi(2v) - 8\varphi(v), \varepsilon\right) \geq \frac{\varepsilon}{\varepsilon + \tau(v, v, 0, \ldots, 0)} \tag{5.61}$$

for all $v \in E$ and $\varepsilon > 0$. Using (F3), we have

$$N_\beta\left(\frac{\varphi(2v)}{2^2} - \varphi(v), \frac{\varepsilon}{2^{3\beta}}\right) \geq \frac{\varepsilon}{\varepsilon + \tau(v, v, 0, \ldots, 0)} \tag{5.62}$$

for all $v \in E$ and $\varepsilon > 0$. Now, we define a set $X = \{n : E \to F \mid n(0) = 0\}$ and define a generalized metric on X as

$$d(p, n) = \inf\left\{\delta \in [0, \infty] \,\middle|\, N_\beta\left(p(v) - n(v), \delta\varepsilon\right) \geq \frac{\varepsilon}{\varepsilon + \tau(v, v, 0, \ldots, 0)},\ v \in E, \varepsilon \rangle 0\right\},$$

then (X, d) is complete.

We define an operator $M : X \to X$ by $(Mn)(v) = \frac{1}{2^2} n(2v)$ for all $v \in E$. Now, we show that M is strictly contractive on X. For $p, n \in X$, let $d(p, n) = \sigma$. Then

$$N_\beta\left(p(v) - n(v), \sigma\varepsilon\right) \geq \frac{\varepsilon}{\varepsilon + \tau(v, v, 0, \ldots, 0)}$$

for all $v \in E$ and $\varepsilon > 0$. Hence,

$$N_\beta\left((Mp)(v) - (Mn)(v), \sigma L\varepsilon\right) = N_\beta\left(\frac{p(2v)}{2^2} - \frac{n(2v)}{2^2}, \sigma L\varepsilon\right)$$

$$= N_\beta\left(p(2v) - n(v), 2^{2\beta}\sigma L\varepsilon\right)$$

$$\geq \frac{2^{2\beta} L\varepsilon}{2^{2\beta} L\varepsilon + \tau(2v,2v,0,\ldots,0)}$$

$$\geq \frac{\varepsilon}{\varepsilon + \tau(v,v,0,\ldots,0)}$$

for all $v \in E$ and $\varepsilon > 0$. Therefore, $d(Mp,Mn) \leq \sigma L$. This implies $d(Mp,Mn) \leq Ld(p,n)$ for all $p,n \in X$. Next, we prove that $d(M\varphi,\varphi) < \infty$. From inequality (5.62), we obtain

$$N_\beta\left((M\varphi)(v) - \varphi(v), \frac{\varepsilon}{\varepsilon + \tau(v,v,0,\ldots,0)}\right)$$

for all $v \in E$ and $\varepsilon > 0$. That is,

$$d(M\varphi,\varphi) \leq \frac{1}{2^\beta(2^{2\beta})} < \infty. \tag{5.63}$$

Now, it follows from fixed point theorem that there exists $Q_2 : E \to F$ with $Q_2(0) = 0$, a fixed point of M. That is $M(2v) = 2^2 M(v)$, so that $M^i\varphi \to Q_2$, namely

$$Q_2(v) = \lim_{i\to\infty} \frac{\varphi(2^i v)}{2^{2i}}$$

for all $v \in E$. Again, using fixed point theorem and inequality (5.63), we have

$$d(\varphi,Q_2) \leq \frac{1}{1-L} d(M\varphi,\varphi) \leq \frac{1}{2^\beta(2^{2\beta})} \frac{1}{1-L}$$

for all $v \in E$. So that

$$N_\beta\left(\varphi(v) - Q_2(v), \frac{\varepsilon}{2^\beta(2^{2\beta})(1-L)}\right) \geq \frac{\varepsilon}{\varepsilon + \tau(v,v,0,\ldots,0)},$$

Then the inequality (5.60) holds for every $v \in E, \varepsilon > 0$. From inequality (5.58) and (5.59), we get

$$N_\beta\left(\frac{1}{2^{2i}} \nabla\varphi(2^i v_1, 2^i v_2, \ldots, 2^i v_m), \frac{\varepsilon}{2^{2i\beta}}\right) \geq \frac{\varepsilon}{\varepsilon + \tau(2^i v_1, 2^i v_2, \ldots, 2^i v_m)}$$

for all $v_1, v_2, \ldots, v_m \in E, \varepsilon > 0$ and $i \in \mathbb{N}$. From equation (5.58), we obtain

$$N_\beta\left(\frac{1}{2^{2i}} \nabla\varphi(2^i v_1, 2^i v_2, \ldots, 2^i v_m), \varepsilon\right) \geq \frac{2^{2i\beta}\varepsilon}{2^{2i\beta}\varepsilon + 2^{2i\beta}\tau(v_1, v_2, \ldots, v_m)} \geq \frac{\varepsilon}{\varepsilon + L^n\tau(v_1, v_2, \ldots, v_m)}$$

for all $v_1, v_2, \ldots, v_m \in E$ and $\varepsilon > 0$. Since $\lim_{i\to\infty} \frac{\varepsilon}{\varepsilon + L^i \tau(v_1, v_2, \ldots, v_m)} = 1$ for all $v_1, v_2, \ldots, v_m \in E$, $\varepsilon > 0$, we have

$$Q_2\left(\sum_{a=1}^{m} v_a\right) + \sum_{b=1}^{m} Q_2\left(-v_b + \sum_{\substack{a=1;\\ a\neq b}}^{m} v_i\right) = (m-3)\sum_{a=1}^{m} Q_2(v_a + v_b)$$

$$-\left(m^2 - 5m + 4\right)\sum_{a=1}^{m}\left[\frac{Q_2(v_a) - Q_2(-v_a)}{2}\right]$$

$$+\left(-m^2 + 5m - 2\right)\sum_{a=1}^{m}\left[\frac{Q_2(v_a) + Q_2(-v_a)}{2}\right]$$

for all $v_1, v_2, \ldots, v_m \in E, \varepsilon > 0$.

Suppose the inequality (5.60) is satisfied by another quadratic mapping $Q_2' : E \to F$ besides Q_2. Q_2' satisfying $Q_2'(v) = \frac{1}{2^2} Q_2'(2v)$ $\forall\ v \in E$ and a fixed point Q_2' of M.

From equation (5.60) and the definition of d, we obtain that

$$d\left(\varphi, Q_2'\right) \leq \frac{1}{2^{2\beta}(1-L)} < \infty,$$

then

$$d\left(M\varphi, Q_2'\right) \leq d\left(M\varphi, \varphi\right) + d\left(\varphi, Q_2'\right) \leq \frac{1}{2^{3\beta}} + \frac{1}{2^{3\beta}(1-L)} < \infty.$$

Hence, this proves that the function Q_2 is unique. The proof is now completed.

Corollary 5.5

Let $w \geq 0, \gamma \in R^+$ with $0 < \gamma < 2$, a real vector space E with β-norm $\|\cdot\|_\beta$ with $0 < \beta \leq 1$. If a mapping $\varphi : E \to F$ is even, which fulfilling $\varphi(0) = 0$ and

$$N_\beta\left(\nabla\varphi(v_1, v_2, \ldots, v_m), \varepsilon\right) \geq \frac{\varepsilon}{\varepsilon + w\left(\sum_{i=1}^{m} v_{i\beta}^{\gamma}\right)} \tag{5.64}$$

for all $v_1, v_2, \ldots, v_m \in E, \varepsilon > 0$, then there exists a unique quadratic mapping $Q_2 : E \to F$ such that

$$N_\beta\left(\varphi(v) - Q_2(v), \varepsilon\right) \geq \frac{2^\beta\left(2^{2\beta} - 2^{\gamma\beta}\right)\varepsilon}{2^\beta\left(2^{2\beta} - 2^{\gamma\beta}\right)\varepsilon + 2wv_\beta^\gamma}$$

for all $v \in E$ and $\varepsilon > 0$.

Theorem 5.8

Let a mapping $\tau : E^m \to [0,\infty)$, there is a Lipschitz constant L, $0 < L < 1$, such that

$$\tau\left(\frac{v_1}{2^i}, \frac{v_2}{2^i}, \dots, \frac{v_m}{2^i}\right) \le \frac{L}{2^{2i\beta}} \tau(v_1, v_2, \dots, v_m) \tag{5.65}$$

for all $v_1, v_2, \dots, v_m \in E$. Suppose an even function $\varphi : E \to F$ satisfying $\varphi(0) = 0$ and the inequality (5.59), then there exists a unique quadratic mapping $Q_2 : E \to F$ satisfying

$$N_\beta(\varphi(v) - Q_2(v), \varepsilon) \ge \frac{2^{3\beta}(L^{-1} - 1)\varepsilon}{2^{3\beta}(L^{-1} - 1)\varepsilon + \tau(v, v, 0, \dots, 0)} \tag{5.66}$$

for all $v \in E$ and $\varepsilon > 0$.

Corollary 5.6

Let $w \ge 0, \gamma \in R^+$ with $\gamma > 2$ and a real vector space E with β-norm $\|\cdot\|_\beta$ with $0 < \beta \le 1$. If an even mapping $\varphi : E \to F$ satisfying $\varphi(0) = 0$ and the inequality (5.64), then there exists a unique quadratic mapping $Q_2 : E \to F$ that satisfies

$$N_\beta(\varphi(v) - Q_2(v), \varepsilon) \ge \frac{2^\beta(2^{\gamma\beta} - 2^{2\beta})\varepsilon}{2^\beta(2^{\gamma\beta} - 2^{2\beta})\varepsilon + 2wv_\beta^\gamma}$$

for all $v \in E$ and $\varepsilon > 0$.

5.4.2 Stability Results When φ is Odd

Theorem 5.9

Let a function $\tau : E^m \to [0,\infty)$, there is a Lipschitz constant L, $0 < L < 1$, such that

$$\tau(2^i v_1, 2^i v_2, \dots, 2^i v_m) \le 2^{i\beta} L \tau(v_1, v_2, \dots, v_m) \tag{5.67}$$

for all $v_1, v_2, \dots, v_m \in E$ and $\varepsilon > 0$. Assume that an odd mapping $\varphi : E \to F$ satisfies $\varphi(0) = 0$ and

$$N_\beta(\nabla\varphi(v_1, v_2, \dots, v_m), \varepsilon) \ge \frac{\varepsilon}{\varepsilon + \tau(v_1, v_2, \dots, v_m)} \tag{5.68}$$

for all $v_1, v_2, \dots, v_m \in E$ and $\varepsilon > 0$. Then there exists a unique mapping $A_1 : E \to F$ that satisfies

$$N_\beta(\varphi(v)-A_1(v),\varepsilon)\geq\frac{2^{2\beta}(1-L)\varepsilon}{2^{2\beta}(1-L)\varepsilon+\tau(v,v,0,\ldots,0)} \tag{5.69}$$

for all $v\in E$ and $\varepsilon>0$.

Proof: Replacing $(v_1,v_2,\ldots,v_m)$ by $(v,v,0,\ldots,0)$ in equation (5.68), we have

$$N_\beta(2\varphi(2v)-4\varphi(v),\varepsilon)\geq\frac{\varepsilon}{\varepsilon+\tau(v,v,0,\ldots,0)} \tag{5.70}$$

for all $v\in E$ and $\varepsilon>0$. Using (F3), we have

$$N_\beta\left(\frac{\varphi(2v)}{2}-\varphi(v),\frac{\varepsilon}{2^{2\beta}}\right)\geq\frac{\varepsilon}{\varepsilon+\tau(v,v,0,\ldots,0)} \tag{5.71}$$

for all $v\in E$ and $\varepsilon>0$. Now, we define a set $X=\{n:E\to F\mid n(0)=0\}$ and define a generalized metric on X as

$$d(p,n)=\inf\left\{\delta\in[0,\infty]\,\middle|\,N_\beta(p(v)-n(v),\delta\varepsilon)\geq\frac{\varepsilon}{\varepsilon+\tau(v,v,0,\ldots,0)},\ v\in E,\varepsilon\rangle 0\right\},$$

then (X,d) is complete.

We define an operator $M:X\to X$ by $(Mn)(v)=\frac{1}{2}n(2v)$ for all $v\in E$. Now, we show that M is strictly contractive on X. For $p,n\in X$, let $d(p,n)=\sigma$. Then

$$N_\beta(p(v)-n(v),\sigma\varepsilon)\geq\frac{\varepsilon}{\varepsilon+\tau(v,v,0,\ldots,0)}$$

for all $v\in E$ and $\varepsilon>0$. Hence,

$$N_\beta((Mp)(v)-(Mn)(v),\sigma L\varepsilon)=N_\beta\left(\frac{p(2v)}{2}-\frac{n(2v)}{2},\sigma L\varepsilon\right)$$

$$=N_\beta(p(2v)-n(v),2^\beta\sigma L\varepsilon)$$

$$\geq\frac{2^\beta L\varepsilon}{2^\beta L\varepsilon+\tau(2v,2v,0,\ldots,0)}$$

$$\geq\frac{\varepsilon}{\varepsilon+\tau(v,v,0,\ldots,0)}$$

for all $v\in E$ and $\varepsilon>0$. Therefore, $d(Mp,Mn)\leq\sigma L$. This implies $d(Mp,Mn)\leq.$ $Ld(p,n)$ $\forall\ p,n\in X$ Next, we prove that $d(M\varphi,\varphi)<\infty$. From inequality (5.71), we obtain

$$N_\beta\left((M\varphi)(v)-\varphi(v),\frac{\varepsilon}{\varepsilon+\tau(v,v,0,\ldots,0)}\right)$$

for all $v\in E$ and $\varepsilon>0$. That is,

$$d(M\varphi,\varphi)\le\frac{1}{2^\beta(2^\beta)}<\infty. \tag{5.72}$$

Now, it follows from fixed point theorem that there exists a function $A_1:E\to F$ with $A_1(0)=0$, which is a fixed point of M. That is, $M(2v)=2M(v)$, so that $M^i\varphi\to A_1$, namely

$$A_1(v)=\lim_{i\to\infty}\frac{\varphi(2^i v)}{2^i}$$

for all $v\in E$. Again, using fixed point theorem and inequality (5.72), we have

$$d(\varphi,A_1)\le\frac{1}{1-L}d(M\varphi,\varphi)\le\frac{1}{2^\beta(2^\beta)}\frac{1}{1-L}$$

for all $v\in E$. So that

$$N_\beta\left(\varphi(v)-A_1(v),\frac{\varepsilon}{2^\beta(2^\beta)(1-L)}\right)\ge\frac{\varepsilon}{\varepsilon+\tau(v,v,0,\ldots,0)},$$

Then, the inequality (5.69) holds for every $v\in E,\varepsilon>0$. From inequality (5.67) and (5.68), we get

$$N_\beta\left(\frac{1}{2^i}\nabla\varphi(2^iv_1,2^iv_2,\ldots,2^iv_m),\frac{\varepsilon}{2^{i\beta}}\right)\ge\frac{\varepsilon}{\varepsilon+\tau(2^iv_1,2^iv_2,\ldots,2^iv_m)}$$

for all $v_1,v_2,\ldots,v_m\in E,\varepsilon>0$, and $i\in\mathbb{N}$. From equation (5.67), we obtain

$$N_\beta\left(\frac{1}{2^i}\nabla\varphi(2^iv_1,2^iv_2,\ldots,2^iv_m),\varepsilon\right)\ge\frac{2^{i\beta}\varepsilon}{2^{i\beta}\varepsilon+2^{2i\beta}\tau(v_1,v_2,\ldots,v_m)}\ge$$

$$\frac{\varepsilon}{\varepsilon+L^n\tau(v_1,v_2,\ldots,v_m)}\quad\forall\ v_1,v_2,\ldots,v_m\in E,\ \varepsilon>0$$

Since $\lim_{i\to\infty}\frac{\varepsilon}{\varepsilon+L^i\tau(v_1,v_2,\ldots,v_m)}=1$ for all $v_1,v_2,\ldots,v_m\in E$, $\varepsilon>0$, we have

$$A_1\left(\sum_{a=1}^{m} v_a\right)+\sum_{b=1}^{m} A_1\left(-v_b+\sum_{\substack{a=1;\\ a\neq b}}^{m} v_i\right)=(m-3)\sum_{a=1}^{m} A_1(v_a+v_b)$$

$$-\left(m^2-5m+4\right)\sum_{a=1}^{m}\left[\frac{A_1(v_a)-A_1(-v_a)}{2}\right]$$

$$+\left(-m^2+5m-2\right)\sum_{a=1}^{m}\left[\frac{A_1(v_a)+A_1(-v_a)}{2}\right]$$

for all $v_1, v_2, \ldots, v_m \in E, \varepsilon > 0$. Assume that the inequality (5.69) is also satisfied with another additive mapping $A_1^{'}: E \to F$ besides A_1. $A_1^{'}$ satisfying $A_1^{'}(v)=\frac{1}{2}A_1^{'}(2v)\ \forall\ v \in E$, and a fixed point $A_1^{'}$ of M.

From equation (5.69) and the definition of d, we obtain that

$$d\left(\varphi, A_1^{'}\right) \leq \frac{1}{2^{\beta}(1-L)} < \infty,$$

then

$$d\left(M\varphi, A_1^{'}\right) \leq d(M\varphi, \varphi) + d\left(\varphi, A_1^{'}\right) \leq \frac{1}{2^{2\beta}} + \frac{1}{2^{2\beta}(1-L)} < \infty.$$

Hence, this shows that the function A_1 is unique. The proof is now completed.

Corollary 5.7

Let $w \geq 0, \gamma \in R^{+}$ with $0 < \gamma < 1$. If an odd mapping $\varphi: E \to F$ satisfying $\varphi(0)=0$ and

$$N_{\beta}\left(\nabla\varphi(v_1, v_2, \ldots, v_m), \varepsilon\right) \geq \frac{\varepsilon}{\varepsilon + w\left(\sum_{i=1}^{m} v_{i\beta}^{\gamma}\right)} \tag{5.73}$$

for all $v_1, v_2, \ldots, v_m \in E, \varepsilon > 0$, then there exists a unique additive function $A_1: E \to F$ that satisfies

$$N_{\beta}\left(\varphi(v) - A_1(v), \varepsilon\right) \geq \frac{2^{\beta}\left(2^{\beta}-2^{\gamma\beta}\right)\varepsilon}{2^{\beta}\left(2^{\beta}-2^{\gamma\beta}\right)\varepsilon + 2wv_{\beta}^{\gamma}}$$

for all $v \in E$ and $\varepsilon > 0$.

Theorem 5.10

Let a mapping $\tau : E^m \to [0,\infty)$, there is a Lipschitz constant L, $0 < L < 1$, such that

$$\tau\left(\frac{v_1}{2^i}, \frac{v_2}{2^i}, \ldots, \frac{v_m}{2^i}\right) \le \frac{L}{2^{i\beta}} \tau(v_1, v_2, \ldots, v_m) \tag{5.74}$$

for all $v_1, v_2, \ldots, v_m \in E$ and $\varepsilon > 0$. Suppose an odd mapping $\varphi : E \to F$ satisfies $\varphi(0) = 0$ and the inequality (5.68), then there exists a unique additive function $A_1 : E \to F$ that satisfies

$$N_\beta(\varphi(v) - A_1(v), \varepsilon) \ge \frac{2^{2\beta}(L^{-1} - 1)\varepsilon}{2^{2\beta}(L^{-1} - 1)\varepsilon + \tau(v, v, 0, \ldots, 0)} \tag{5.75}$$

for all $v \in E$ and $\varepsilon > 0$.

Corollary 5.8

Let $w \ge 0, \gamma \in R^+$ with $\gamma > 1$. If an odd mapping $\varphi : E \to F$ satisfies $\varphi(0) = 0$ and satisfies the inequality (5.73), then there is a unique additive mapping $A_1 : E \to F$ that satisfies

$$N_\beta(\varphi(v) - A_1(v), \varepsilon) \ge \frac{2^\beta(2^{\gamma\beta} - 2^\beta)\varepsilon}{2^\beta(2^{\gamma\beta} - 2^\beta)\varepsilon + 2wv_\beta^\gamma}$$

for all $v \in E$ and $\varepsilon > 0$.

REFERENCES

Alanazi, A.M., G. Muhiuddin, K. Tamilvanan, E.N. Alenze, A. Ebaid, and K. Loganathan. 2020. Fuzzy stability results of finite variable additive functional equation: direct and fixed point methods. *Mathematics* 8: 1050. Doi: 10.3390/math8071050.

Bag, T. and S.K. Samanta 2003. Finite dimensional fuzzy normed linear spaces. *J. Fuzzy Math.* 11(3): 687–705.

Cadariu, L. and V. Radu. 2008. Fixed point methods for the generalized stability of functional equations in a single variable. *Fixed Point Theory Appl.*: 1–5 (Art. ID 749392).

Cadariu, L. and V. Radu. 2004. On the stability of the Cauchy functional equation: a fixed point approach. *Grazer Math Ber.*: 43–52.

Cheng, S.C. and J.N. Moderson 1994. Fuzzy linear operators and fuzzy normed linear spaces. *Bull. Calcutta Math. Soc.* 86: 429–436.

Felbin, C. 1992. Finite-dimensional fuzzy normed linear space. *Fuzzy Sets Syst.* 48: 143–154.

Hyers, D.H. 1941. On the stability of the linear functional equation. *Proc. Nat. Acad. Sci., U.S.A.* 27: 222–224.

Isac, G. and Th.M. Rassias. 1996. Stability of Ã-additive mappings applications to nonlinear analysis. *Inter. J. Math. Math. Sci.* 19: 219–228. Doi:10.1155/S0161171296000324.

Kaleva, O. and S. Seikkala. 1984. On fuzzy metric spaces. *Fuzzy Sets Syst.* 12: 215–229.

Katsaras, A.K. 1981. Fuzzy topological vector spaces I. *Fuzzy Sets Syst.* 6: 85–95.

Katsaras, A.K. and D.B. Liu. 1977. Fuzzy vector spaces and fuzzy topological vector spaces. *J. Math. Anal. Appl.* 58: 135–146.

Lee, S.J., C. Park, and R. Saadati. 2012. Orthogonal stability of an additive-quartic functional equation with the fixed point alternative. *J. Inequal. Appl.* 83. Doi: 10.1186/1029-242X-2012-83.

Lowen, R. 1976. Fuzzy topological spaces and fuzzy compactness. *J. Math. Anal. Appl.* 56: 621–633.

Park, C. 2011. Orthogonal stability of an additive-quadratic functional equation. *J. Inequal. Appl.* 66. Doi: 10.1186/1687-1812-2011-66.

Park, C., K. Tamilvanan, B. Noori, M.B. Moghimi, and A. Najati. 2020. Fuzzy normed spaces and stability of a generalized quadratic functional equation. *AIMS Math.* 5(6): 7161–7174.

Radu, V. 2003. The fixed point alternative and the stability of functional equations. *Fixed Point Theor.* 4: 91–96.

Tamilvanan, K., G. Balasubramanian, and A. Charles Sagayaraj. 2020a. Finite dimensional even-quadratic functional equation and its Ulam-Hyers stability. *AIP Conf. Proc.* 2261: 030002. Doi: 10.1063/5.0016865.

Tamilvanan, K., J.R. Lee, and C. Park. 2020b. Hyers-Ulam stability of a finite variable mixed type quadratic-additive functional equation in quasi-Banach spaces. *AIMS Math.* 5(6): 5993–6005.

Ulam, S.M. 1960. *Problems in Modern Mathematics, Rend. Chap.VI*, Wiley, New York.

Yang, X., G. Shen, G. Liu, and L. Chang. 2015. The Hyers-Ulam-Rassias stability of the quartic functional equation in fuzzy β-normed spaces. *J. Inequal. Appl.* 2015: 342. Doi: 10.1186/s13660-015-0863-5.

6 Tauberian Theorems for Intuitionistic Fuzzy Normed Spaces

Vakeel A. Khan and Mobeen Ahmad
Aligarh Muslim University

CONTENTS

6.1 INTRODUCTION

In 1965, Zadeh introduced the notion of fuzzy sets theory as the generalization of classical set theory. As per the theme of concept of fuzzy sets are concerned, the researchers have utilized the idea. In recent years, this theory has acquired more attention from authors in a wide range of disciplines. The purpose is to apply the notion of fuzziness to individual works with distinct situation from theoretical to practical in almost all technology and sciences. Researchers have obtained varied and uncountable applications of this theory in fields such as pattern recognition, artificial intelligence, computer science, finance, decision theory and stock market, nuclear science, agriculture, geography, biomedicine, statistics, etc. In pure mathematics, various authors have used fuzziness in a wide range of applications. Matloka (1986) initiated the notion of ordinary convergence of a sequence of fuzzy real numbers and studied convergent and bounded sequences of fuzzy numbers and some of their properties. Nanda (1989) investigated some basic stuff for these sequences and showed that the set of all convergent sequences of fuzzy real numbers forms a complete metric space. Savas (2000) studied on sequence of fuzzy numbers. Esi and Acikgoz (2014) analyzed almost A-statistical convergence of fuzzy numbers. Moreover, Nuray and Savas (1995) extended the notion of convergence of the sequence of fuzzy real numbers to the notion of statistical convergence. Alaba and Norahun (2019) studied fuzzy Ideals and fuzzy filters of pseudo-complemented semilattices.

As an extension of fuzzy sets theory, Atanassov (1986) introduced the concepts of intuitionistic fuzzy sets which is a generalization of fuzzy sets. Followed by intuitionistic fuzzy sets, Park (2004) introduced intuitionistic fuzzy metric, and Saadati and Park (2006) further analyzed the concept of intuitionistic fuzzy norm (in short,

IFN) spaces. In 2008, Lael and Nourouzi defined the intuitionistic fuzzy norm space and examined some fundamental theorems (Lael and Nourouzi, 2008). The statistical convergence for sequences in IFN spaces was defined and obtained some important results (Karakus et al. 2008). In 2009, Mursaleen and Mohiuddine examined the statistical convergence, lacunary statistical convergence and ideal convergence of double sequences in IFN spaces (Mursaleen and Mohiuddine, 2009a, 2009b, Mursaleen et al. 2010). Debnath and Sen analyzed statistical convergence and calculus for functions having values in an intuitionistic fuzzy n-normed linear space (Debnath and Sen, 2011, 2014). In 2019, the ideal convergence of double sequence in IFN space defined by bounded linear operator and ideal convergence Fibonacci difference sequences in IFN spaces were studied by Khan et al. (2019a, 2019b).

Móricz (1994) has proved some Tauberian theorem for Cesaro summable double sequences. Statistical extension of some classical Tauberian theorem was studied by Fridy and Khan (2000). Canak et al. analyzed Tauberian theorem for (C, 1, 1) summable double sequences of fuzzy numbers (Canak et al. 2017), and further, Talo and Yavuz investigated Tauberian theorems for single sequences in IFN spaces (Talo and Yavuz 2020).

6.2 PRELIMINARIES

In this section, we recall some basic notions, definitions and lemma that are required for the following sections.

Definition 6.1

(Lael and Nourouzi 2008) The triplicate (U,ϕ,ψ) is known as an IFN space if U is a real vector space and ϕ, ψ are fuzzy sets on $U \times R$ satisfying conditions for every u, $v \in U$ and α, $\beta \in$ R:

i. $\varphi(u,\alpha)=0$ for all $\alpha < 0$
ii. $\varphi(u,\alpha)=1$ for all $\alpha > 0$ if and only if $u = \theta$
iii. $\varphi(cu,\alpha)=\varphi\left(u,\frac{\alpha}{c}\right)$ for all $\alpha > 0$ and c $\neq 0$,
iv. $\varphi(u+v,\ \alpha+\beta) \geq \min\{\varphi(u,\alpha),\varphi(v,\ \beta)\}$
v. $\lim_{\alpha\to\infty}\varphi(u,\alpha)=1$ and $\lim_{\alpha\to 0}\varphi(u,\alpha)=0$
vi. $\psi(u,\alpha)=1$ for all $\alpha < 0$
vii. $\psi(u,\alpha)=0$ for all $\alpha < 0$ if and only if $u = \theta$
viii. $\psi(cu,\alpha)=\psi\left(u,\frac{\alpha}{c}\right)$ for all $\alpha > 0$ and c $\neq 0$,
ix. max $\{\varphi(u,\alpha),\varphi(v,\beta)\} \geq \varphi(u+v,\alpha+\beta)$
x. $\lim_{a\to\infty}\psi(u,\alpha)=0$ and $\lim_{a\to 0}\psi(u,\alpha)=1$

In such case, (ϕ, ψ) is said to be an intuitionistic fuzzy norm (IFN).

Example 6.1

(Lael and Nourouzi 2008) Let $(U, \| \ . \ \|)$ be a norm space and ϕ_0 and ψ_0 be fuzzy sets on $U^2 \times (0,\infty)$ defines as follows:

$$\underset{0}{\varphi}(u,\alpha) = \begin{cases} 0, & \alpha \le 0, \\ \dfrac{\alpha}{\alpha + \| u \|}, & \alpha > 0, \end{cases}$$

$$\psi_0(u,\alpha) = \begin{cases} 1, & \alpha \le 0, \\ \dfrac{\| u \|}{\alpha + \| u \|}, & \alpha > 0. \end{cases}$$

Then (ϕ_0, ψ_0) is an IFN on U.

Theorem 6.1

(Lael and Nourouzi 2008) Let (U, ϕ, ψ) be an IFN space. Assume further that

$$\phi(u,\alpha) > 0 \text{ for all } \alpha > 0 \text{ implies} u = 0 \tag{6.1}$$

Define

$$\| u \|_\lambda = \inf\{\alpha > 0 : \varphi(u,\alpha) > \lambda \text{ and } \psi(u,\alpha) < 1 - \lambda\}$$

where $\lambda \in (0,1)$. Then $\{\| u \|_\lambda : \lambda \in (0,1)\}$ is an ascending family of norms on U.

These norms are called λ –norms on U corresponding to the IF-norm (ϕ, ψ) on U.

Definition 6.2

(Lael and Nourouzi 2008) A sequence (u_m) in an IFN space (U, ϕ, ψ) is called convergent to ξ if for all $\alpha > 0$ and for any $\varepsilon \in (0,1)$, there exists $m_0 \in \mathrm{N}$ in such a way that $\phi(u_m - \xi, \alpha) > 1 - \varepsilon$ and $\psi(u_m - \xi, \alpha) < \varepsilon$ for all $m > m_0$.

Definition 6.3

(Lael and Nourouzi 2008) A sequence (u_m) in an IFN space (U, ϕ, ψ) is called Cauchy if for all $\alpha > 0$ and for any $\varepsilon \in (0,1)$, there exists $m_0 \in \mathrm{N}$ in such a way that $\phi(u_m - u_n, \alpha) > 1 - \varepsilon$ and $\psi(u_m - u_n, \alpha) < \varepsilon$ for all $n, m > m_0$.

Definition 6.4

(Grabiec 1988). A sequence (u_m) in an IFN space (U, ϕ, ψ) is G-Cauchy if $\lim_{m\to\infty}\varphi\left(u_{m+p}-u_m,\alpha\right)=1$ and $\lim_{m\to\infty}\psi\left(u_{m+p}-u_m,\alpha\right)=0$ for each $\alpha>0\, p\in\mathrm{N}$. This definition is equivalent to $\lim_{m\to\infty}\psi\left(u_{m+1}-u_m,\alpha\right)=1$ and $\lim_{m\to\infty}\psi\left(u_{m+1}-u_m,\alpha\right)=0$ for all $\alpha > 0$.

Remark 6.1

(Grabiec 1988). Every convergent sequence is Cauchy, and every Cauchy sequence is G-Cauchy in an intuitionistic fuzzy normed space.

Next, we recall the concepts of boundedness and boundedness in an IFN space.

Definition 6.5

(Efe and Alaca 2007). Let (U, ϕ, ψ) be an IFN space, and B be any subset of U.

i. A is called bounded if there exist some $r \in (0,1)$ and $\alpha_0 > 0$ such that ϕ $(u, \alpha_0) > 1 - r$ and $\psi\,(u, \alpha_0) < r$ for every $u \in B$.
ii. B is called q-bounded if $\lim\limits_{\alpha\to\infty} \phi\, B\,(\alpha) = 1$, and $\lim\limits_{\alpha\to\infty} \psi\, B\,(\alpha) = 0$ where

$$\varphi_B(\alpha)=\inf\{\mu(u,\alpha):u\in B\},$$

$$\psi_B(\alpha)=\sup\{\nu(u,\alpha):u\in B\},$$

It can be easily seen that the sequence (u_k) in an IFN space (U, ϕ, ψ) is bounded if and only if there exists some $\alpha_0 > 0$ and $r \in (0,1)$ such that $\phi\,(u_m, \alpha_0) > 1-r$ and ψ $(u_m, \alpha_0) < r$ for every positive integer m and q-bounded if and only if

$$\lim_{a\to\infty}\inf_{m\in N}\varphi\left(u_m,\alpha\right)=1 \text{ and } \lim_{a\to\infty}\sup_{m\in N}\psi\left(u_m,\alpha\right)=0 \tag{6.2}$$

Definition 6.6

(Lael and Nourouzi 2008) Let (u_m) be a sequence in IFN space (U, ϕ, ψ).

Then

$$\sigma_m=\frac{1}{m+1}\sum_{j=0}^{m}u_j$$

Where σ_m is the arithmetic mean of the sequence (u_m). The sequence is called Cesaro summable to $\xi \in U$ if

$$\lim_{m\to\infty}\sigma_m=\xi.$$

Talo and Basar (2018) had given the following lemma:

Lemma 6.1

(Talo and Basar 2018) We have the following statement:

i. Suppose $\gamma > 1$, for each $m \in N \setminus \{0\}$ with $m \geq \frac{3\gamma - 1}{\gamma(\gamma - 1)}$, one has

$$\frac{\gamma}{\gamma - 1} < \frac{\gamma_m + 1}{\gamma_m - m} < \frac{2\gamma}{1 - \gamma}$$

ii. Suppose $0 < \gamma < 1$, for each $m \in N \setminus \{0\}$ with $m > \frac{1}{\gamma}$, one has

$$0 < \frac{\gamma_m + 1}{m - \gamma_m} < \frac{2\gamma}{1 - \gamma}.$$

Definition 6.7

(Móricz 1994) A double sequence (u_{ij}: j, k = 0,1…) converges to a finite number ξ if $u_{jk} \rightarrow \xi$ as j, $k \rightarrow \infty$, independently of one another.

Defining the means (C, 1,1) of (u_{jk}) by

$$\sigma_{mn} := \frac{1}{(m+1)(n+1)} \sum_{j=0}^{m} \sum_{k=0}^{n} u_{jk} \,(m, n = 0, 1 \ldots.)$$

One says that sequence (u_{jk}) is summable (C, 1,1) to finite limit ξ if the mean $\sigma_{\text{mn-}}$ converges to ξ.

Lemma 6.2

(Móricz 1994)

i. If $\gamma > 1, \gamma_m > m$ and $\gamma_n > n$, then

$$u_{mm} - \sigma_{mn} = \Gamma_{mn} - \frac{1}{(\gamma_m - m)(\gamma_n - n)} \sum_{k=m+1}^{\gamma_m} \sum_{l=n+1}^{\gamma_n} (u_{mn} - u_{kl}). \tag{6.3}$$

Where

$$\Gamma_{mn} = \frac{(\gamma_m + 1)(\gamma_n + 1)}{(\gamma_m - m)(\gamma_n - n)} \left(\sigma_{\gamma m, \gamma n} - \sigma_{\gamma n, n} - \sigma_{m, \lambda n} + \sigma_{mn}\right)$$

$$+ \frac{(\gamma_m + 1)}{(\gamma_m - m)} \left(\sigma_{\gamma m, n} - \sigma_{mn}\right) + \frac{(\gamma_n + 1)}{(\gamma_n - n)} \left(\sigma_{m, \gamma_n} - \sigma_{mn}\right) \tag{6.4}$$

ii. If $0 < \gamma < 1$, $\gamma_m < m$ and $\gamma_n < n$, then

$$u_{mn} - \sigma_{mn} = \Lambda_{mn} + \frac{1}{(m-\gamma_m)(n-\gamma_n)} \sum_{k=\gamma_m+1}^{m} \sum_{l=\gamma_n+1}^{n} (u_{mn} - u_{kl}). \tag{6.5}$$

Where

$$\Lambda_{mn} = \frac{(\gamma_m+1)(\gamma_n+1)}{(m-\gamma_m)(n-\gamma_n)}\left(\sigma_{\gamma m,n} - \sigma_{m,\gamma n} + \sigma_{\gamma m,\gamma_n}\right)$$

$$+\frac{(\gamma_m+1)}{(m-\gamma_m)}\left(\sigma_{mn} - \sigma_{\gamma m,n}\right) + \frac{(\gamma_n+1)}{(n-\gamma_n)}\left(\sigma_{mn} - \sigma_{m,\gamma n}\right) \tag{6.6}$$

6.3 MAIN RESULTS

Definition 6.8

Let (U, ϕ, ψ) be an IFN space. A sequence $(u_{mn}) \in U$ is called slowly oscillating if for each $\alpha > 0$,

$$\sup_{\gamma>1} \liminf_{m,n\to\infty} \min_{m<j<\gamma_m, n<k<\gamma_n} \phi\left(u_{jk} - u_{mn,}\alpha\right) = 1 \tag{6.7}$$

and

$$\inf_{\gamma>1} \limsup_{m,n\to\infty} \max_{m<j<\gamma_m, n<k<\gamma_n} \psi\left(u_{jk} - u_{mn}, \alpha\right) = 0 \tag{6.8}$$

where γ_m and γ_n denote the integer part of the product γ_m and γ_n, respectively.

Sup $\gamma_{>1}$ and inf $\gamma_{>1}$ can be replaced by $\lim_{\gamma} \rightarrow_{1+.}$ In other words, a double sequence (u_{jk}) is said to be slowly oscillating if and only if for all $\alpha > 0$ and for all $\varepsilon \in (0,1)$ there exists $\gamma > 1$ and $m_0 \in N$, depending on α and ε in such a way that

$$\phi\left(u_{jk} - u_{mn,}a\right) > 1 - \in \text{and} \psi\left(u_{jk} - u_{mn,}a\right) \tag{6.9}$$

whenever $m_0 \leq m < j \leq \gamma_m$ and $n_0 \leq n < k \leq \gamma_n$.

Theorem 6.2

Let (U, ϕ, ψ) be an IFN space and (u_{mn}) be a sequence in U. If the sequence (x_{mn}) is covergent to $\xi \in U$, then (u_{mn}) is Cesáro summable to ξ.

Proof: Suppose the sequence (x_{mn}) is a convergent to $\xi \in U$. Select $\alpha > 0$ and given $\varepsilon > 0$.

i. Then there exists $m_0 \in N$ such that $\phi\left(u_{mn} - \xi, \frac{\alpha}{2}\right) < \varepsilon$ for m, $n > m_0$.

ii. There exists $m_1 \in N$ such that

$$\phi\left(\sum_{j=0}^{m_0}\sum_{k=0}^{m_0}(u_{jk}-\xi),\frac{(n+1)(m+1)\alpha}{2}\right)>1-\varepsilon \text{ and}$$

$\phi\left(\sum_{j=0}^{m_0}\sum_{k=0}^{n_0}(u_{jk}-\xi),\frac{(n+1)(m+1)\alpha}{2}\right)<\varepsilon$ for $m, n > m_1$; therefore, whenever $m > \max\{m_0, m_1\}$ and $n > \max\{n_0, n_1\}$, we have

$$\lim_{m,n\to\infty}\varphi\left(\sum_{j=0}^{m_0}\sum_{k=0}^{n_0}(u_{jk}-\xi),\frac{(n+1)(m+1)\alpha}{2}\right)=1$$

and

$$\lim_{m,n\to\infty}\psi\left(\sum_{j=0}^{m_0}\sum_{k=0}^{n_0}(u_{jk}-\xi),\frac{(n+1)(m+1)\alpha}{2}\right)=0.$$

Therefore,

$$\phi\left(\frac{1}{(n+1)(m+1)}\sum_{j=0}^{m_0}\sum_{k=0}^{n_0}u_{jk}-\xi,a\right)=\phi\left(\frac{1}{(n+1)(m+1)}\sum_{j=0}^{m}\sum_{k=0}^{n}(u_{jk}-\xi),\ \alpha\right)$$

$$=\phi\left(\sum_{j=0}^{m}\sum_{k=0}^{n}(u_{jk}-\xi),(n+1)(m+1)\alpha\right)$$

$$\geq \min\left\{\phi\left(\sum_{j=0}^{m_0}\sum_{k=0}^{n_0}(u_{jk}-\xi),\frac{(n+1)(m+1)\alpha}{2}\right),\right.$$

$$\left.\phi\left(\sum_{j=m_0+1}^{m}\ \sum_{k=m_0+1}^{n}(u_{jk}-\xi),\frac{(n+1)(m+1)\alpha}{2}\right)\right\}$$

$$\geq \min\left\{\phi\left(\sum_{j=0}^{m_0}\sum_{k=0}^{n_0}(u_{jk}-\xi),\frac{(n+1)(m+1)\alpha}{2}\right)\right.$$

$$\left.\phi\left(\sum_{j=0}^{m}\sum_{k=0}^{n}(u_{jk}-\xi),\frac{(n-n_0)(m-m_0)\alpha}{2}\right)\right\}$$

$$\geq \min\left\{\phi\left(\sum_{j=0}^{m_0}\sum_{k=0}^{n_0}(u_{jk}-\xi),\frac{(n+1)(m+1)\alpha}{2}\right),\right.$$

$$\varphi\left(u_{(m_0+1)(n_0+1)}-\xi,\frac{a}{2}\right),\varphi\left(u_{(m_0+2)(n_0+2)}-\xi,\frac{a}{2}\right),$$

$$\varphi\left(u_{(m_0+3)(n_0+3)}-\xi,\frac{a}{2}\right),\dots,\varphi\left(u_{mn}-\xi,\frac{a}{2}\right)\Big\}$$

$$>1-\varepsilon$$

and

$$\psi\left(\frac{1}{(n+1(m+1))}\sum_{j=0}^{m_0}\sum_{k=0}^{n_0}(u_{jk}-\xi),\alpha\right)<\max\left\{\psi\left(\sum_{j=0}^{m_0}\sum_{k=0}^{n_0}(u_{jk}-\xi),\frac{(n+1)(m+1)\alpha}{2}\right),\right.$$

$$\psi\left(u_{(m_0+1)(n_0+1)}-\xi,\frac{\alpha}{2}\right),\varphi\left(u_{(m_0+2)(n_0+2)}-\xi,\frac{\alpha}{2}\right),$$

$$\left.\psi\left(u_{(m_0+3)(n_0+3)}-\xi,\frac{a}{2}\right),\dots,\psi\left(u_{mn}-\xi,\frac{a}{2}\right)\right\}$$

$$<\varepsilon.$$

Whenever $m > \max\{m_0, m_1\}$ and $n > \max\{n_0, n_1\}$.

This proved that Cesáro summability method is regular in IFN space.

Theorem 6.3

Let (U, ϕ, ψ) be an IFN space. A double sequence (u_{mn}) is Cesáro summable to $\xi \in U$, then it converges to ξ if and only if for all $\alpha > 0$

$$\sup_{\gamma>1}\liminf_{m,n\to\infty}\varphi\left(\frac{1}{(\gamma_m-m)(\gamma_n-n)}\right)\sum_{j=m+1}^{\gamma_m}\sum_{k=n+1}^{\gamma_n}(u_{jk}-u_{mn},\alpha)=1 \tag{6.10}$$

and

$$\inf_{\gamma>1}\lim_{m,n\to\infty}\sup\psi\left(\frac{1}{(\gamma_m-m)(\gamma_n-n)}\right)\sum_{j=m+1}^{\gamma_m}\sum_{k=n+1}^{\gamma_n}(u_{jk}-u_{mn},\alpha)=0. \tag{6.11}$$

Proof: Suppose (u_{mn}) be a sequence in (U, ϕ, ψ) and Cesára summable to á $\xi \in U$.

Necessary condition: Suppose (u_{mn}) converges to ξ, fix $\alpha > 0$. For any $\gamma > 1$, one has

$$u_{mn}-\sigma_{mn}=\Gamma_{mn}-\frac{1}{(\gamma_m-m)(\gamma_n-n)}\sum_{k=m+1}^{\gamma_m}\sum_{l=n+1}^{\gamma_n}(u_{kl}-u_{mn}). \tag{6.12}$$

With the help of Lemma 6.1, for $m > (3\gamma - 1)/\gamma(\gamma - 1)$, one obtains

$$\varphi(\Gamma_{mn},\alpha) = \varphi\left(\frac{(\gamma_m - 1)(\gamma_n - 1)}{(\gamma_m - m)(\gamma_n - n)}(\sigma_{\gamma m,\gamma n} - \sigma_{m,n} - \sigma_{m,\gamma n} + \sigma_{mn},\alpha)\right)$$

$$+\varphi\left(\frac{(\gamma_m + 1)}{(\gamma_m - m)}(\sigma_{\gamma_m,n} - \sigma_{mn},\alpha) + \varphi\left(\frac{(\gamma_n + 1)}{(\gamma_n - n)}\right)(\sigma_{m,\gamma_n} - \sigma_{mn},\alpha)\right)$$

$$= \varphi\left((\sigma_{\gamma m,\gamma n} - \sigma_{m,n} - \sigma_{m,\gamma n} + \sigma_{mn}),\frac{(\gamma_m - m)(\gamma_n - n)}{(\gamma_m + 1)(\gamma_n + 1)}\alpha\right)$$

$$+\phi\left((\sigma_{\gamma_m,n} - \sigma_{mn})\frac{(\gamma_m - m)}{(\gamma_m + 1)}\alpha\right) + \phi\left((\sigma_{m,\gamma_n} - \sigma_{mn}),\frac{(\gamma_n - n)}{(\gamma_n + 1)}\alpha\right)$$

$$\geq \phi\left((\sigma_{\gamma m,\gamma n} - \sigma_{\gamma m,n} - \sigma_{m,\gamma_n} + \sigma_{mn}),\frac{\alpha}{\frac{4\gamma^2}{(\gamma - 1)^2}}\right)$$

$$+\varphi\left((\sigma_{\gamma m,n} - \sigma_{mn})\frac{\alpha}{\frac{4\gamma^2}{(\gamma - 1)^2}}\right) + \varphi\left((\sigma_{m,\gamma n} - \sigma_{mn}),\frac{\alpha}{\frac{4\gamma^2}{(\gamma - 1)^2}}\right)$$

and

$$\psi(\Gamma_{mn},\alpha) = \psi\left(\frac{(\gamma_m - 1)(\gamma_n - 1)}{(\gamma_m - m)(\gamma_n - n)}(\sigma_{\gamma m,\gamma n} - \sigma_{m,n} - \sigma_{m,\gamma n} + \sigma_{mn},\alpha)\right)$$

$$+\psi\left(\frac{(\gamma_m + 1)}{(\gamma_m - m)}(\sigma_{\gamma_m,n} - \sigma_{mn},\alpha) + \psi\left(\frac{(\gamma_n + 1)}{(\gamma_n - n)}\right)(\sigma_{m,\gamma_n} - \sigma_{mn},\alpha)\right)$$

$$= \psi\left((\sigma_{\gamma m,\gamma n} - \sigma_{m,n} - \sigma_{m,\gamma n} + \sigma_{mn}),\frac{(\gamma_m - m)(\gamma_n - n)}{(\gamma_m + 1)(\gamma_n + 1)}\alpha\right)$$

$$+\psi\left((\sigma_{\gamma_m,n} - \sigma_{mn})\frac{(\gamma_m - m)}{(\gamma_m + 1)}\alpha\right) + \psi\left((\sigma_{m,\gamma_n} - \sigma_{mn}),\frac{(\gamma_n - n)}{(\gamma_n + 1)}\alpha\right)$$

$$\leq \psi\left((\sigma_{\gamma m,\gamma n} - \sigma_{\gamma m,n} - \sigma_{m,\gamma_n} + \sigma_{mn}),\frac{\alpha}{\frac{4\gamma^2}{(\gamma - 1)^2}}\right)$$

$$+\psi\left((\sigma_{\gamma m,n}-\sigma_{mn})\frac{\alpha}{\frac{4\gamma^2}{(\gamma-1)^2}}\right)+\psi\left((\sigma_{m,\gamma n}-\sigma_{mn}),\frac{\alpha}{\frac{4\gamma^2}{(\gamma-1)^2}}\right)$$

Since (σ_{mn}) is a Cauchy Sequence, $\lim_{m,n\to\infty}\phi(\Gamma_{mn,}\ \alpha)=1$ and $\lim_{m,n\to\infty}\psi(\Gamma_{mn,}\ \alpha)=0$ which means that

$$\lim_{m,n\to\infty}\phi\left(\frac{1}{(\gamma_m-m)(\gamma_n-n)}\right)\sum_{j=n+1}^{\gamma_n}\sum_{k=m+1}^{\delta_m}(u_{jk}-u_{nm},\alpha)=1 \tag{6.13}$$

and

$$\lim_{n,m\to\infty}\psi\left(\frac{1}{(\gamma_m-m)(\gamma_n-n)}\right)\sum_{j=n+1}^{\gamma_n}\sum_{k=m+1}^{\delta_m}(u_{jk}-u_{nm},\alpha)=0. \tag{6.14}$$

Therefore, equations (6.10) and (6.11) are proved.

Sufficient Condition: Assuming that equations (6.10) and (6.11) are satisfied, choose $\alpha > 0$. Therefore, for given $\varepsilon > 0$, one obtains:

i. Let $m_0 \in$ N and there exists $\gamma > 0$ in such a way that

$$\phi\left(\frac{1}{(\gamma_m-m)(\gamma_n-n)}\right)\sum_{j=m+1}^{\gamma m}\sum_{k=n+1}^{\gamma_n}\left(u_{jk}-u_{mn},\frac{\alpha}{3}\right)>1-\varepsilon$$

and

$$\psi\left(\frac{1}{(\gamma_m-m)(\gamma_n-n)}\right)\sum_{j=m+1}^{\gamma m}\sum_{k=n+1}^{\gamma_n}\left(u_{jk}-u_{mn},\frac{\alpha}{3}\right)<\varepsilon$$

for all $m, n > m_0$.

ii. There exists $m_1 \in N$ in such a way that

$$\phi\left(\sigma_{mn}-\xi,\frac{\alpha}{3}\right)>1-\varepsilon$$

and

$$\psi\left(\sigma_{mn}-\xi,\frac{\alpha}{3}\right)<\varepsilon$$

for $m, n > m_1$.

iii. There exists $m_2 \in N$ in such a way that

$$\varphi\left(\frac{(\gamma_m+1)(\gamma_n+1)}{(\gamma_m-m)(\gamma_n-n)}\right)\left(\sigma_{\gamma_m\gamma_n}-\sigma_{mn},\frac{\alpha}{3}\right)>1-\varepsilon$$

and

$$\psi\left(\frac{(\gamma_m+1)(\gamma_n+1)}{(\gamma_m-m)(\gamma_n-n)}\right)\left(\sigma_{\gamma_m\gamma_n}-\sigma_{mn},\frac{\alpha}{3}\right),<\varepsilon$$

As we know the fact that $\frac{(\gamma_m+1)}{(\gamma_m-m)}\frac{(\gamma_n+1)}{(\gamma_n-n)}\left(\sigma_{\gamma_m\gamma_n}-\sigma_{mn}\right)\to 0$. Therefore, one obtains

$$\varphi\left(u_{mn}-\xi,a\right)=\varphi\left(u_{mn}-\sigma_{mn}+\sigma_{mn}-\xi,\alpha\right)$$

$$=\varphi\left(\frac{(\gamma_m+1)}{(\gamma_m-m)}\frac{(\gamma_n+1)}{(\gamma_n-n)}\right)\left(\sigma_{\gamma_m\gamma_n}-\sigma_{mn}\right)$$

$$-\frac{1}{(\gamma_m-m)(\gamma_n-n)}\sum_{j=m+1}^{\gamma_m}\sum_{k=n+1}^{\gamma_n}\left(u_{jk}-u_{mn}\right)+\sigma_{mn}-\xi,a)$$

$$\geq \min\left\{\phi\left(\frac{(\gamma_m+1)}{(\gamma_m-m)}\frac{(\gamma_n+1)}{(\gamma_n-n)}\left(\sigma_{\gamma_m\gamma_n}-\sigma_{mn},\frac{\alpha}{3}\right)\right),\right.$$

$$\left.\phi\frac{1}{(\gamma_m-m)(\gamma_n-n)}\sum_{j=m+1}^{\gamma_m}\sum_{k=n+1}^{\gamma_n}\left(u_{jk}-u_{mn},\frac{\alpha}{3}\right)+\phi\left(\sigma_{mn}-\xi,\frac{\alpha}{3}\right)\right\}$$

$$>1-\varepsilon$$

and

$$\psi\left(u_{mn}-\xi,\alpha\right)\geq \max\left\{\psi\left(\frac{(\gamma_m+1)}{(\gamma_m-m)}\frac{(\gamma_n+1)}{(\gamma_n-n)}(\sigma_{\gamma_m\gamma_n}-\sigma_{mn},\frac{\alpha}{3}\right),\right.$$

$$\left.\psi\frac{1}{(\gamma_m-m)(\gamma_n-n)}\sum_{j=m+1}^{\gamma_m}\sum_{k=n+1}^{\gamma_n}\left(u_{jk}-u_{mn},\frac{\alpha}{3}\right)+\psi\left(\sigma_{mn}-\xi,\frac{\alpha}{3}\right)\right\}$$

$$<\varepsilon$$

for $m > \max\{m_0, m_1, m_2\}$ and $n > \max\{n_0, n_1, n_2\}$ which completes the proof.

Theorem 6.4

Let (U, ϕ, ψ) be an IFN space and (u_{mn}) be a sequence in U. If (u_{mn}) is Cesáro summable to $\xi \in U$, then it converges to ξ if and only if for all $\alpha > 0$.

$$\sup_{0<\gamma<1} \liminf_{m,n\to\infty} \phi\left(\frac{1}{(\gamma_m - m)(\gamma_n - n)}\right) \sum_{j=m+1}^{\gamma_m} \sum_{k=n+1}^{\gamma_n} (u_{jk} - u_{mn}, \alpha) = 1$$

and

$$\inf_{0<\gamma<1} \limsup_{m,n\to\infty} \psi\left(\frac{1}{(\gamma_m - m)(\gamma_n - n)}\right) \sum_{j=m+1}^{\gamma_m} \sum_{k=n+1}^{\gamma_n} (u_{jk} - u_{nm}, \alpha) = 0.$$

Proof: The proof of the theorem is similar to the proof of Theorem 6.3; therefore, it is omitted.

Theorem 6.5

Let (u_{mn}) be a sequence in IFN space (U, ϕ, ψ). For $\alpha > 0$, conditions (6.7) and (6.8) of slow oscillation are equivalent to

$$\sup_{0<\gamma<1} \liminf_{m,n\to\infty} \max_{\gamma_m<j\leq m, \gamma_n<k\leq n} \phi(u_{jk} - u_{mn}, \alpha) = 1 \tag{6.15}$$

and

$$\inf_{0<\gamma<1} \limsup_{m,n\to\infty} \max_{\gamma_m<j\leq m, \gamma_n<k\leq n} \varphi(u_{jk} - u_{mn}, \alpha) = 0 \tag{6.16}$$

respectively. $\text{Sup}_{0<\gamma<1}$ in (6.15) and $\inf_{0<\lambda<1}$ in (6.16) can be replaced by $\lim_{\lambda\to 1} -$.

Proof: We will prove that equation (6.7) and equation (6.15) are equivalent. Assume $\alpha > 0$ and suppose that

$$g(\gamma) \liminf_{m,n\to\infty} \min_{m<j[\gamma_m], n<k\leq[\gamma_n]} \varphi(u_{jk} - u_{mn}, \alpha) \text{ and}$$

$$h\left(\frac{1}{\gamma}\right) = \liminf_{jk\to\infty} \min_{\left[\frac{j}{\gamma}\right]<m\leq j, [\gamma m], \left[\frac{k}{\gamma}\right]<n\leq k} \phi(u_{jk} - u_{mn}, \alpha)$$

where $\gamma > 1$. Therefore, for each $\gamma > 1$, there exists increasing sequences (m_p) and (n_p) in such a way that

$$g(\gamma) = \liminf_{p\to\infty} \min_{m_p<j\leq[\gamma m_p], n_p<k\leq[\gamma n_p]} \phi(u_{jk} - u_{m_p n_p}, \alpha)$$

Also, there exists sequences $(j_p) \in \left(m, \lfloor \gamma m_p \rfloor\right]$ and $(k_p) \in \left(n, \lfloor \gamma n_p \rfloor\right]$ in such a way that

$$\min_{mp<k\leq[\gamma m_p]} \phi\left(u_{jk} - u_{m_p n_p}, \alpha\right) = \phi\left(u_{jk} - u_{m_p n_p}, \alpha\right)$$

Here, we note that $j_p \in \left(m, \lfloor \gamma m_p \rfloor\right]$ implies $m_p \in \left(\left\lfloor \frac{jp}{\gamma} \right\rfloor, j_p\right)$ and $k_p\left(m, \lfloor \gamma m_p \rfloor\right]$ implies $m_p \in \left(\left\lfloor \frac{kp}{\gamma} \right\rfloor, k_p\right)$ Here, one obtains

$$h\left(\frac{1}{\gamma}\right) = \liminf_{j,k\to\infty} \min_{\left\lfloor \frac{j}{\gamma} \right\rfloor < m \leq j, \left\lfloor \frac{k}{\gamma} \right\rfloor < n \leq k} \phi\left(u_{jk} - u_{mn}, \alpha\right)$$

$$\leq \lim_{p\to\infty} \min_{\left\lfloor \frac{jp}{\gamma} \right\rfloor < m \leq j, \left\lfloor \frac{kp}{\gamma} \right\rfloor < m \leq kp} \phi\left(u_{j_p kp} - u_{mn}, \alpha\right)$$

$$\leq \lim_{p\to\infty} \phi\left(u_{j_p k_p} - u_{n_p m_p}, \alpha\right)$$

$$= \lim_{p\to\infty} \inf_{m_p<j\leq[\gamma m_p], n_p<k\leq[\gamma n_p]} \min \left(u_{jk} - u_{n_p m_p}, \alpha\right)$$

$$= g(\gamma).$$

On the contrary, if we change the roles of $h\left(\frac{1}{\gamma}\right)$ and $g(\gamma)$, and putting the same process, one obtains $h\left(\frac{1}{\gamma}\right) \geq g(\gamma)$. Therefore, $\gamma > 1$, one has $g(\gamma) = h\left(\frac{1}{\gamma}\right)$, which implies that equation (6.7) and equation (6.15) are equivalent.

In the same way, one can show that equations (6.8) and (6.16) are equivalent.

Theorem 6.6

Let $(U, \|\cdot\|)$ be a normed space and suppose that (U, ϕ_0, ϕ_0) be IFN space, where ϕ_0 and ϕ_0 satisfy the conditions given by Example 6.1. A double sequence (u_{mn}) is slowly oscillating in $(U, \|\cdot\|)$ if and only if (u_{mn}) is slowly oscillating in (U, ϕ_0, ϕ_0).

Proof: Suppose that double sequence (u_{mn}) is slowly oscillating in (U, ϕ_0, ϕ_0), then there exists $\gamma > 1$, for given $\varepsilon \in (0, \frac{1}{2})$ and $m_0, n_0 \in N$ in such a way that

$$\varphi_0\left(u_{jk}-u_{mn},1\right)=\frac{1}{1+\| u_{jk}-u_{mn}\|}>1-\varepsilon$$ Whenever $m_0 \le m \le j \le \gamma_m$ and $n_0 \le n \le k \le \gamma_m$

Therefore, one obtains

$\| u_{jk}-u_{mn} \| < \dfrac{\varepsilon}{1-\varepsilon} < 2\varepsilon$ whenever $m_0 \le m \le j \le \gamma_m$ and $n_0 \le n \le k \le \gamma_m$. This shows that (u_{mn}) is slowly oscillating in $(U, \|\cdot\|)$.

Conversely, suppose that a $\left(u_{jk}\right)$ be slowly oscillating in $(U,\|.\|)$. For given $\alpha > 0$ and $\varepsilon \in (0, 1)$, one defines $\varepsilon_0 = \dfrac{\alpha\varepsilon}{1-\varepsilon}$. Then there exist $\gamma > 1$ and $m_0, n_0 \in N$ in such a way that $\| u_{jk}-u_{mn} \| < \varepsilon_0$ whenever $m_0 \le m \le j \le \gamma_m$ and $n_0 \le n \le k \le \gamma_m$. Therefore,

$$\phi_0\left(u_{jk}-u_{mn},\alpha\right)=\frac{\alpha}{\alpha+\| \varepsilon_0-u_{mn}\|}>\frac{\alpha}{\alpha+\varepsilon_0}=1-\varepsilon$$

and

$$\psi_0\left(u_{jk}-u_{mn},\alpha\right)=1-\phi_0\left(u_{jk}-u_{mn},\alpha\right)<\varepsilon,$$

whenever $m_0 \le m \le j \le \gamma_m$ and $n_0 \le n \le k \le \gamma_m$. This means that (u_{mn}) is slowly oscillating in $\left(\mathrm{U}, \varphi_0, \psi_0\right)$.

It can be easily seen that

$$\text{Cauchy sequence} \Rightarrow \text{Slow obcillation sequence} \Rightarrow \text{G}-\text{Cauchy sequence}$$

But it cannot be reverted in general.

Example 6.2

Consider IFN space (R, ϕ_0, ψ_0) where ϕ_0 and ψ_0 as in Example 6.1. By Theorem 6.6, the double sequence $\left(u_{jk}\right)$ defined by $u_{jk}=\sum_{j,k=1}^{m,n}\frac{1}{\sqrt{jk}}$ is G-Cauchy sequence but is not slowly oscillating sequence, and the sequence $\left(v_{jk}\right)$ defined by $u_{jk} = \sum_{j,k=1}^{m,n}\frac{1}{jk}$ is slowly oscillating sequence but is not Cauchy sequence.

Theorem 6.7

Let $\left(U,\phi,\psi\right)$ be an IFN space satisfying condition (6.1) and $\left(u_{mn}\right)$ be a sequence in U. The double sequence $\left(u_{mn}\right)$ is slowly oscillating in $\left(U,\phi,\psi\right)$ if and only if $\left(u_{mn}\right)$ is slowly oscillating in $\left(U,\|.\|_\delta\right)$ for each $\delta \in (0.1)$.

Proof: Suppose that $\delta \in (0,1)$ and $\beta > 0$. Let the sequence (u_{mn}) is slowly oscillating in (U, ϕ, ψ). Then, for $\varepsilon = 1-\delta$ there exist $\gamma > 1$ and m_0, $n_0 \in N$ so that $m_0 \le m < j \le \gamma_m$ and $n_0 \le n < k \le \gamma_{n,}$ one obtains

$$\phi\left(u_{jk}-u_{mn},\beta\right)>1-\varepsilon \text{ and } \psi\left(u_{jk}-u_{mn},\beta\right)<\varepsilon$$

and

$$\| u_{jk}-u_{mn} \|_{\delta}=\inf\left\{\alpha>0:\phi\left(u_{jk}-u_{mn},\alpha\right)>\delta \text{ and } \psi\left(u_{jk}-u_{mn},\alpha\right)<1-\delta\right\}<\beta$$

Hence, sequence (u_{mn}) is slowly oscillating in $(U, \|\cdot\| \delta)$.

Conversely, select $\delta \in (0,1)$ and let (u_{mn}) be slowly oscillating in $(U, \|\cdot\| \delta)$. Then, for $\beta > 0$, there exists $\gamma > 1$ and m_0, $n_0 \in N$ in such a way that

$$\| u_{jk}-u_{mn} \|_{\delta}=\inf\{\alpha>0:\varphi\left(u_{jk}-u_{mn},\alpha\right)>\delta \text{ and } \psi\left(u_{jk}-u_{mn},\alpha\right)<1-\delta\}<\beta$$

whenever $m_0 \le m < j \le \gamma_m$ and $n_0 \le n < k \le \gamma_{n.}$

Therefore, $\phi\left(u_{jk}-u_{mn},\beta\right)>\delta$ and $\psi\left(u_{jk}-u_{mn},\beta\right)<1-\delta$ whenever $m_0 \le m < j \le \gamma_m$ and $n_0 \le n < k \le \gamma_{n.}$

Since β and δ were arbitrary, (u_{mn}) is slowly oscillating in (U, ϕ, ψ).

Now, we prove that slow oscillating condition is a Tauberian condition for Cesáro summability method in an IFN space.

Theorem 6.8

Let (u_{mn}) be a slowly oscillating sequence in an IFN space (U, ϕ, ψ). If (u_{mn}) is slowly oscillating, then equations (6.10) and (6.11) are satisfied.

Proof: Suppose that (u_{mn}) be a slowly oscillating sequence in an IFN space (U, ϕ, ψ). If (u_{mn}). Choose $\alpha > 0$. Therefore, for given $\varepsilon \in (0,1)$ there exist $\gamma > 1$ and $m_0, n_0 \in N$ such a way that

$$\phi\left(u_{jk}-x_{mn},\alpha\right)>1-\varepsilon \text{ and } \psi\left(u_{jk}-u_{mn},\alpha\right)<\varepsilon \text{ whenever } m_0 \le m<j\le\gamma_m \text{ and}$$

$$n_0 \le n<k\le\gamma_n$$

Therefore, one has

$$\phi\left(\frac{1}{(\gamma_m-m)(\gamma_n-n)}\sum_{j=m+1}^{\gamma_m}\sum_{k=n+1}^{\gamma_n}\left(u_{jk}-u_{mn},\alpha\right)\right)$$

$$=\phi\sum_{k=m+1}^{\gamma_m}\sum_{k=n+1}^{\gamma_n}\left(u_{jk}-u_{mn},(\gamma_m-m)(\gamma_n-n)\alpha\right)$$

$$\ge\min\left\{\phi\left(u_{(m+1)(n+1)}-u_{mn},\alpha\right),\phi\left(u_{(m+2)(n+2)}-u_{mn},\alpha\right),\ldots,\phi\left(u_{\gamma_m\gamma_n}-u_{mn},\alpha\right)\right\}$$

$$>1-\varepsilon$$

and

$$\psi\left(\frac{1}{(\gamma_m - m)(\gamma_n - n)}\sum_{j=m+1}^{\gamma_m}\sum_{k=n+1}^{\gamma_n}(u_{jk} - u_{mn}, a)\right)$$

$$\psi\left(\sum_{j=m+1}^{\gamma_m}\sum_{k=n+1}^{\gamma}(u_{jk} - u_{mn}, (\gamma_m - m)(\gamma_n - n)\alpha)\right)$$

$$\geq \max\left\{\psi\left(u_{(m+1)(n+1)} - u_{mn}, \alpha\right), \psi\left(u_{(m+2)(n+2)} - u_{mn}, \alpha\right), \ldots, \psi\left(u_{\gamma_m \gamma_n} - u_{mn}, \alpha\right)\right\}$$

whenever $m_0 \leq j \leq \gamma_m$ and $n_0 \leq n \leq k \leq \gamma_n$, which completes the proof.

We obtain the following conclusion from the Theorems 6.2 and 6.8, which is as follows:

Theorem 6.9

Let the double sequence (u_{mn}) in IFN space be (U, ϕ, ψ). If sequence (u_{mn}) is Cesáro summable to $\xi \in U$ and slowly oscillating, then sequence (u_{mn}) converges to ξ.

Now, we show the comparison theorem between q-boundedness of the sequences and the notion of slow oscillation in intuitionistic fuzzy normed spaces.

Theorem 6.10

Let (u_{mn}) be the double sequence in (U, ϕ, ψ). If $mn\left(u_{mn} - u_{(m-1)(n-1)}\right)$ is q-bounded, then sequence (u_{mn}) is slowly oscillating.

Proof: For a given $\varepsilon \in (0,1)$ and using equation (6.2), there exists $N \in > 0$ in such a way that

$$\alpha > N, \Rightarrow \inf_{m,n \in n} \phi\left(mn\left(u_{mn} - u_{(m-1)(n-1)}\right), \alpha\right) > 1 - \varepsilon$$

and

$$\sup_{m,n \in n} \psi\left(mn\left(u_{mn} - u_{(m-1)(n-1)}\right), \alpha\right) < \varepsilon$$

For all $\alpha > 0$, select $\gamma < 1 + \frac{\alpha}{N_\varepsilon}$, then $m_0 < m < j \leq \gamma_m$ and $n_0 < n < k \leq \gamma_n$

$$\phi\left(u_{jk} - u_{mn,}\alpha\right) = \phi\left(\sum_{p=m+1}^{j}\sum_{q=n+1}^{k}\left(u_{pq} - u_{(p-1)(q-1)}, \alpha\right)\right)$$

$$\min_{m+1\le p\le j, n+1\le q\le k} \phi\left(\left(u_{pq} - u_{(p-1)(q-1)}, \alpha\right)\right)$$

$$= \min_{m+1\le p\le j, n+1\le q\le k} \phi\left(pq\left(u_{pq} - u_{(p-1)(q-1)}, \frac{mn\alpha}{(j-m)(k-n)}\right)\right)$$

$$\ge \min_{m+1\le p\le j, n+1\le q\le k} \phi\left(pq\left(u_{pq} - u_{(p-1)(q-1)}, \frac{\alpha}{\left(\frac{j}{m}-1\right)\left(\frac{k}{n}-1\right)}\right)\right)$$

$$\ge \min_{m+1\le p\le j, n+1\le q\le k} \phi\left(pq\left(u_{pq} - u_{(p-1)(q-1)}, \frac{\alpha}{(\gamma-1)^2}\right)\right)$$

$$\ge \inf_{m,n\in N} \phi\left(mn\left(u_{mn} - u_{(m-1)(n-1)}\right), \frac{\alpha}{(\gamma-1)^2}\right)$$

$$> 1-\varepsilon$$

and in the same manner

$$\phi\left(u_{jk} - u_{mn}, \alpha\right) < \sup \psi\left(mn\left(u_{mn} - u_{(m-1)(n-1)}\right), \frac{\alpha}{(\gamma-1)^2}\right) < \varepsilon.$$

This implies that sequence (u_{mn}) is slowly oscillating.

6.4 CONCLUSION

In this chapter, we provide an introduction to the summability theory and Tauberian theory for double sequence in intuitionistic fuzzy normed spaces, with some basic definitions of convergence. We also defined the notion of slowly oscillating double sequence in intuitionistic fuzzy normed space and proved some important results. These results are very useful and interesting in many areas of mathematics and engineering.

REFERENCES

Alaba, B. A., and W. Z. Norahun. 2019. Fuzzy ideals and fuzzy filters of pseudocomplemented semilattices. *Advances in Fuzzy Systems*: 1–13.

Atanassov, K. T. 1986. Intuitionistic fuzzy sets. *Fuzzy Sets System* 20: 87–96.

Canak, I., U. Totur, and Z. Onder. 2017. A Tauberian theorem for (C, 1, 1) summable double sequences of fuzzy numbers. *Iranian Journal of Fuzzy Systems* 14(1): 61–75.

Debnath, P., and M. Sen, 2011. Statistical convergence in intuitionistic fuzzy n-normed linear spaces. *Fuzzy Information and Engineering* 3(3): 259–273.

Debnath, P., and M. Sen. 2014. Some results of calculus for functions having values in an intuitionistic fuzzy n-normed linear space. *Journal of Intelligent & Fuzzy Systems* 26(6): 2983–2991.

Efe, H. and C. Alaca. 2007. Compact and bounded sets in intuitionistic fuzzy metric spaces. *Demonstratio Mathematica* 40(2): 449–456.

Esi A., and M. Acikgoz. 2014. On almost A-statistical convergence of fuzzy numbers. *Acta Scientiarum Technology* 36(1): 129–133.

Fridy, J. A., and M. K. Khan. 2000. Statistical extensions for some classical Tauberian theorems. *Proceedings of the American Mathematical Society* 128(8): 2347–2355.

Grabiec, M. 1988. Fixed points in fuzzy metric spaces. *Fuzzy Sets and Systems* 27(3): 385–389.

Karakus, S., K. Demirci, and O. Duman. 2008. Statistical convergence on intuitionistic fuzzy normed spaces. *Chaos, Solitons & Fractals* 35(4): 763–769.

Khan, V. A., M. Ahmad, H. Fatima, and M. F. Khan. 2019a On some results in intuitionistic fuzzy ideal convergence double sequence spaces. *Advances in Difference Equations* 2019: 1–10.

Khan, V. A., E. E. Kara, H. Altaf, and M. Ahmad. 2019b. Intuitionistic fuzzy I-convergent Fibonacci difference sequence spaces. *Journal of Inequalities and Applications* 2019(1): 202.

Lael, F., and K. Nourouzi. 2008. Some results on the IF-normed spaces. *Chaos, Solitons & Fractals* 37(3): 931–939.

Matloka, M. 1986. Sequences of fuzzy numbers. *Busefal* 28(1): 28–37.

Móricz, F. 1994. Tauberian theorems for Cesaro summable double sequences. *Studia Mathematicn* 1(110): 83–96.

Mursaleen, M., and S. A. Mohiuddine. 2009a. On lacunary statistical convergence with respect to the intuitionistic fuzzy normed space. *Journal of Computational and Applied Mathematics* 233(2): 142–149.

Mursaleen, M., and S. A. Mohiuddine. 2009b. Statistical convergence of double sequences in intuitionistic fuzzy normed spaces. *Chaos Solitons & Fractals* 41(5): 2414–2421.

Mursaleen, M., S. A. Mohiuddine and O. H. H. Edely. 2010. On the ideal convergence of double sequences in intuitionistic fuzzy noned spaces. *Computers & Mathematics with Applications* 59(2): 603–611.

Nanda, S. 1989. On sequences of fuzzy numbers. *Fuzzy Sets and Systems* 33(1): 123–126.

Nuray, F., and E. Savaş. 1995. Statistical convergence of sequences of fuzzy numbers. *Mathematical Slovaca* 45(3): 269–273.

Park, J. H. 2004. Intuitionistic fuzzy metric spaces. *Chaos, Solitons & Fractals*, 22(5): 1039–1046.

Saadati, R., and J. H. Park. 2006. On the intuitionistic fuzzy topological spaces. *Chaos, Solutions & Fractals* 27(2): 331–344.

Savas, E. 2000. A note on sequence of fuzzy numbers. *Information Sciences* 124(4): 297–300.

Talo, O., and E. Yavuz. 2020. Cesáro summability of sequences in intuitionistic fuzzy normed spaces and related Tauberian theorems. *Soft Computing*: 1–9.

Talo, O., and F. Basar. 2018. Necessary and sufficient Tauberian conditions for the A method of summability. *Mathematical Journal of Okayama University* 60: 209–219.

Zadeh, L. A. 1965. Fuzzy sets. *Information and Control*, 8(3): 338–353.

7 Picture Fuzzy Soft Matrices

S. Vijayabalaji
University College of Engineering Panruti
(A Constituent College of Anna University)

CONTENTS

7.1 INTRODUCTION

The novel idea of picture fuzzy set (PFS) was initiated by (Cuong 2014), following the concepts of fuzzy set (Zadeh 1965) and intuitionistic fuzzy set (Atanassov 1986). PFS-based models can be applied to situations requiring human opinions involving more than two answers. In a few years, picture fuzzy set obtained the rapid development both in theory and applications. Singh (2014) applied the correlation coefficients for picture fuzzy set in clustering analysis, and Ganie et al. (2020) proposed the correlation coefficients over some existing methods in pattern recognition, medical diagnosis and clustering in PFS. (Thong 2015) applied picture fuzzy set to medical diagnosis and application to health care support systems. Fuzzy cross entropy model (Wei 2016) and weighted cosine function similarity measures between PFSs for multiple attribute decision-making problems were developed by Wei (2017). Some useful extension on PFS can be viewed in the literature (Ashraf et al. 2018a, b, Zeng et al. 2019, Dutta and Ganju 2018).

After the introduction of soft sets (Molodtsov 1999), many theoretical and applied ideas started emerging in the literature. Soft sets are more flexible to describe the uncertainty and imperfection of the objective world. The matrix notion of soft set was proposed by Cagman and Enginoglu (2010). It created a new way for solving MCDM problem. The notion of soft discernibility matrix was proposed in (Feng and Zhou 2014) and nicely applied in MCDM setting. Based on soft discernibility matrix and picture fuzzy set, Yang et al. (2015) proposed a hybrid model termed as adjustable soft discernibility matrix based on picture fuzzy soft set. Khan et al. (2020)

solved decision-making problem using adjustable weighted soft discernibility matrix in generalized picture fuzzy soft setting.

Motivated by the picture fuzzy set and soft set, in this chapter, a new notion namely picture fuzzy soft matrix is introduced. Some types of picture fuzzy soft matrices with some operations on them are also provided. The chapter is concluded with an example on picture fuzzy soft matrix by using the algorithm of VIKOR method.

7.2 PICTURE FUZZY SOFT MATRICES

Inspired by the concept of picture fuzzy relation (Cuong 2014) for two sets, we define picture fuzzy soft relation for a picture fuzzy soft set as follows.

Definition 7.1

A picture fuzzy soft relation (PFSR) is a subset of $U \times E$. It is represented as follows.

$R = \{((x,y), M(x,y), NT(x,y), NV(x,y)) \mid x \in U, y \in E\}$ with $M : U \times E \to [0,1]$, $NT : U \times E \to [0,1]$ and $NV : U \times E \to [0,1]$. M denotes the degree of positive membership, NT being the degree of neutral membership and NV representing the degree of negative membership.

Example 7.1

Let $U = \{u_1, u_2, u_3, u_4\}$ and $E = \{e_1, e_2, e_3\}$. The PFSR on $U \times E$ is defined as follows.

$$R = \{((u_1,e_1),0.7,0.2,0.1),((u_1,e_2),0.4,0.1,0.1),((u_1,e_3),0.5,0.1,0.1),((u_2,e_1),0.6,0.1,0.1),$$
$$((u_2,e_3),0.3,0.1,0.1),((u_3,e_1),0.8,0.1,0.1),((u_3,e_3),0.6,0.2,0.1),((u_4,e_1),0.5,0.2,0.1),$$
$$((u_4,e_2),0.7,0.1,0.1)\}.$$

We proceed to the introduction of picture fuzzy soft matrix using PFSR.

Definition 7.2

Define a function $P : R \to [0,1]$ by

$$P(u,e) = \begin{cases} (M, \text{NT}, \text{NV}), & (u,e) \in R \\ (0,0,0), & (u,e) \notin R. \end{cases}$$

$P(u,e)$ is called picture fuzzy soft matrix (PFSM) or simply P.

We now provide an example for PFSM to validate our definition.

Example 7.2

Consider the following example. The PFSM P is represented as follows.

$$P = \begin{bmatrix} (0.7,0.2,0.1) & (0.4,0.1,0.1) & (0.5,0.1,0.1) \\ (0.6,0.1,0.1) & (0,0,0) & (0.3,01,0.1) \\ (0.8,0.1,0.1) & (0,0,0) & (0.6,0.2,0.1) \\ (0.5,0.2,0.1) & (0.7,0.1,0.1) & (0,0,0) \end{bmatrix}$$

7.3 OPERATIONS ON PFSM

Definition 7.3

A PFSM with a single row is called picture fuzzy soft row matrix (PFSRM).

Example 7.3

Let $U = \{u_1, u_2, u_3, u_4\}$ and $E = \{e_1, e_2, e_3\}$. The PFSR on $U \times E$ is defined as follows.

$$R = \{((u_1,e_1),0.7,0.2,0.1),((u_1,e_2),0.4,0.1,0.1),((u_1,e_3),0.5,0.1,0.1)\}.$$

The PFSRM is

$$P = \begin{bmatrix} (0.7,0.2,0.1) & (0.4,0.1,0.1) & (0.5,0.1,0.1) \end{bmatrix}$$

Definition 7.4

A PFSM with a column row is called picture fuzzy soft column matrix (PFSCM).

Example 7.4

Let $U = \{u_1, u_2, u_3, u_4\}$ and $E = \{e_1, e_2, e_3\}$. The PFSR on $U \times E$ is defined as follows.

$$R = \{((u_1,e_1),0.7,0.2,0.1),((u_2,e_1),0.4,0.1,0.1),((u_3,e_1),0.5,0.1,0.1)\}.$$

The PFSRM is

$$P = \begin{bmatrix} (07,0.2,0.1) \\ (0.4,0.1,0.1) \\ (0.5,01,0.1) \end{bmatrix}$$

$$P = \begin{bmatrix} (0.7,0.2,0.1) & (0.4,0.1,0.1) & (0.5,0.1,0.1) \end{bmatrix}$$

Definition 7.5

A PFSM with equal number of rows and columns is termed as picture fuzzy soft square matrix (PFSSM).

Example 7.5

Let $U=\{u_1,u_2,u_3,u_4\}$ and $E=\{e_1,e_2,e_3,e_4\}$. The PFSR on $U\times E$ is defined as follows.

$$R=\{((u_1,e_1),0.7,0.2,0.1),((u_1,e_2),0.4,0.1,0.1),((u_1,e_3),0.5,0.1,0.1),((u_2,e_1),0.6,0.1,0.1),$$
$$((u_2,e_3),0.3,0.1,0.1),((u_2,e_4),0.3,0.2,0.1),((u_3,e_1),0.8,0.1,0.1),((u_3,e_3),0.6,0.2,0.1),$$
$$((u_3,e_4),0.5,0.3,0.1),((u_4,e_1),0.5,0.2,0.1),$$
$$((u_4,e_2),0.7,0.1,0.1),((u_4,e_4),0.6,0.2,0.1),\}.$$

The PFSSM is

$$P=\begin{bmatrix} (0.7,0.2,0.1) & (0.4,0.1,0.1) & (0.5,0.1,0.1) & (0,0,0) \\ (0.6,0.1,0.1) & (0,0,0) & (0.3,0.1,0.1) & (0.3,0.2,0.1) \\ (0.8,0.1,0.1) & (0,0,0) & (0.6,0.2,0.1) & (0.5,0.3,0.1) \\ (0.5,0.2,0.1) & (0.7,0.1,0.1) & (0,0,0) & (0.6,0.2,0.1) \end{bmatrix}$$

7.4 OPERATIONS ON PFSM

We define several operations on given two picture fuzzy soft matrices as follows.

Definition 7.6

Given two PFS matrices $P=(M_1,\mathrm{NT}_1,\mathrm{NV}_1)$ and $Q=(M_2,\mathrm{NT}_2,\mathrm{NV}_2)$, we define their addition $P+Q$ as follows.

$$P+Q=(\max(M_1,M_2),\min(\mathrm{NT}_1,\mathrm{NT}_2),\min(\mathrm{NV}_1,\mathrm{NV}_2))$$

Example 7.6

Consider $P=\begin{bmatrix} (0.4,0.2,0.1) & (0.5,0.2,0.1) \\ (0.7,0.2,0.1) & (0.6,0.2,0.1) \end{bmatrix}$ and $Q=\begin{bmatrix} (0.6,0.1,0.1) & (0.5,0.2,0.1) \\ (0.3,0.1,0.1) & (0.7,0.1,0.1) \end{bmatrix}$.

Then

$$P+Q=\begin{bmatrix}(0.6,0.1,0.1) & (0.5,0.2,0.1)\\(0.7,0.1,0.1) & (0.7,0.1,0.1)\end{bmatrix}$$

Definition 7.7

Given two PFS matrices $P=(M_1,\mathrm{NT}_1,\mathrm{NV}_1)$ and $Q=(M_2,\mathrm{NT}_2,\mathrm{NV}_2)$, we define their subtraction $P-Q$ as follows.

$$P-Q=(\min(M_1,M_2),\max(\mathrm{NT}_1,\mathrm{NT}_2),\max(\mathrm{NV}_1,\mathrm{NV}_2))$$

Example 7.7

Consider $P=\begin{bmatrix}(0.4,0.2,0.1) & (0.5,0.2,0.1)\\(0.7,0.2,0.1) & (0.6,0.2,0.1)\end{bmatrix}$ and $Q=\begin{bmatrix}(0.6,0.1,0.1) & (0.5,0.2,0.1)\\(0.3,0.1,0.1) & (0.7,0.1,0.1)\end{bmatrix}$.

Then

$$P-Q=\begin{bmatrix}(0.4,0.2,0.1) & (0.5,0.2,0.1)\\(0.3,0.2,0.1) & (0.6,0.2,0.1)\end{bmatrix}$$

Definition 7.8

Given two PFS matrices $P=(M_1,\mathrm{NT}_1,\mathrm{NV}_1)$ and $Q=(M_2,\mathrm{NT}_2,\mathrm{NV}_2)$, we define their product PQ as follows.

$$PQ=(\max(M_1,M_2),\max(\mathrm{NT}_1,\mathrm{NT}_2),\min(\mathrm{NV}_1,\mathrm{NV}_2))$$

Example 7.8

Consider $P=\begin{bmatrix}(0.4,0.2,0.1) & (0.5,0.2,0.1)\\(0.7,0.2,0.1) & (0.6,0.2,0.1)\end{bmatrix}$ and $Q=\begin{bmatrix}(0.6,0.1,0.1) & (0.5,0.2,0.1)\\(0.3,0.1,0.1) & (0.7,0.1,0.1)\end{bmatrix}$.

Then

$$PQ=\begin{bmatrix}(0.6,0.2,0.1) & (0.5,0.2,0.1)\\(0.7,0.2,0.1) & (0.7,0.2,0.1)\end{bmatrix}$$

Definition 7.9

Given two PFS matrices $P=(M_1,\mathrm{NT}_1,\mathrm{NV}_1)$ and $Q=(M_2,\mathrm{NT}_2,\mathrm{NV}_2)$, we define their division P/Q as follows.

$$P/Q=(\min(M_1,M_2),\min(NT_1,\mathrm{NT}_2),\max(NV_1,\mathrm{NV}_2))$$

Example 7.9

Consider $P=\begin{bmatrix}(0.4,0.2,0.1) & (0.5,0.2,0.1)\\ (0.7,0.2,0.1) & (0.6,0.2,0.1)\end{bmatrix}$ and $Q=\begin{bmatrix}(0.6,0.1,0.1) & (0.5,0.2,0.1)\\ (0.3,0.1,0.1) & (0.7,0.1,0.1)\end{bmatrix}$.

Then

$$P/Q=\begin{bmatrix}(0.4,0.1,0.1) & (0.5,0.2,0.1)\\ (0.3,0.1,0.1) & (0.6,0.1,0.1)\end{bmatrix}$$

7.5 RECALL OF VIKOR METHOD

We provide an algorithm to exhibit the application of PFSM.

VIKOR method (Opricovic, 1998) is of the finest method for MCDM problems. This method is one of the compensatory methods, parameters are independent, and qualitative parameters are changed to quantitative attributes. The steps involved in VIKOR method are given precisely for a better understanding.

Algorithm of VIKOR Method

Step 1: Determine the objective; pertinent evaluation attributes the best $m_{ij_{max}}$ and the worst $m_{ij_{min}}$ values of all attributes.

Step 2: Calculate the values of E_i and F_i.

$$E_i=\sum_{j=1}^{m}W_j\left[m_{ij_{\max}}-m_{ij}\right]/\left[m_{ij_{\max}}-m_{ij_{\min}}\right] \tag{7.1}$$

$$F_i=\max\left\{\sum_{j=1}^{m}W_j\left[m_{ij_{\max}}-m_{ij}\right]/\left[m_{ij_{\max}}-m_{ij_{\min}}\right],\ j-1,2,3,\ldots m\right\} \tag{7.2}$$

Step 3: Find P_i. Here

$$P_i=V\left(\frac{E_i-E_{i_{\min}}}{E_{i_{\max}}-E_{i_{\min}}}\right)+(1-V)\left(\frac{F_i-F_{i_{\max}}}{F_{i_{\max}}-F_{i_{\min}}}\right) \tag{7.3}$$

$E_{i_{max}}$ is the maximum value of E_i, and $E_{i_{min}}$ is the minimum value of E_i, $F_{i_{\max}}$ is the maximum value of F_i, and $F_{i_{min}}$ is the minimum value of F_i.

Step 4: Arrange the alternatives in ascending order, according to the values of Pi. Similarly, arrange the alternatives according to the values of E_i and F_i separately. Thus, three ranking lists can be obtained.

Step 5: Rank the alternatives.

The alternatives are ranked as descending in values of E_i, F_i and P_i. The alternative with the lowest amount in attributes is the superior alternative.

We try to apply VIKOR algorithm for PFSM by choosing decision matrix as PFSM.

Example 7.10

A workshop company intends to buy the best CNC lathe model among the models of A_1, A_2 and A_3. The decision attributes are the amount of coolant consumption in liters (C_1) and number of pieces produced per day (C_2) and parameters have equal weights.

We now apply the above algorithm on PFSM to attain the decision goal.

Step 1: Decision matrix of purchasing CNC lathe

$$\begin{matrix} & C_1 & C_2 \\ \begin{matrix} A_1 \\ A_2 \\ A_3 \end{matrix} & \multicolumn{2}{c}{\begin{pmatrix} (0.4,0.2,0.1) & (0.5,0.2,0.1) \\ (0.5,0.2,0.1) & (0.4,0.2,0.1 \\ (0.3,0.2,0.1) & (0.3,0.1,0.2) \end{pmatrix}} \end{matrix}$$

Now we have to find $m_{ij_{\max}}$ and $m_{ij_{\min}}$,

$$m_{1_{\max}} = (0.5,0.2,0.1), m_{1_{\min}} = (0.3,0.2,0.1)$$

$$m_{2_{\max}} = (0.5,0.1,0.1), m_{2_{\min}} = (0.3,0.2,0.2)$$

Step 2: By choosing $w = (0.3,0.2,0.1)$, using (7.1) and (7.2), the values of E_i and F_i are calculated as follows.

$$E_1 = (0.3,0.2,0.1) \times \left[\frac{(0.5,0.2,0.1) - (0.4,0.2,0.1)}{(0.5,0.2,0.1) - (0.3,0.2,0.1)} \right] + (0.3,0.2,0.1)$$

$$\times \left[\frac{(0.5,0.1,0.1) - (0.5,0.2,0.1)}{(0.5,0.1,0.1) - (0.3,0.2,0.2)} \right] = (0.3,0.2,0.1) + (0.3,0.2,0.2) = (0.3,0.2,0.1)$$

$$E_2=(0.3,0.2,0.1)\times\left[\frac{(0.5,0.2,0.1)-(0.5,0.2,0.1)}{(0.5,0.2,0.1)-(0.3,0.2,0.1)}\right]+(0.3,0.2,0.1)$$

$$\times\left[\frac{(0.5,0.1,0.1)-(0.4,0.2,0.1)}{(0.5,0.1,0.1)-(0.3,0.2,0.2)}\right]=(0.3,0.2,0.1)+(0.3,0.2,0.2)=(0.3,0.2,0.1)$$

$$E_3=(0.3,0.2,0.1)\times\left[\frac{(0.5,0.2,0.1)-(0.3,0.2,0.1)}{(0.5,0.2,0.1)-(0.3,0.2,0.1)}\right]+(0.3,0.2,0.1)$$

$$\times\left[\frac{(0.5,0.1,0.1)-(0.3,0.1,0.2)}{(0.5,0.1,0.1)-(0.3,0.2,0.2)}\right]=(0.3,0.3,0.2)+(0.3,0.2,0.2)=(0.3,0.2,0.2)$$

In addition to $F_{i,}$ index values for alternatives are as follows.

$$F_1=\max\left\{\begin{matrix}(0.3,0.2,0.1)\times\left[\frac{(0.5,0.2,0.1)-(0.4,0.2,0.1)}{(0.5,0.2,0.1)-(0.3,0.2,0.1)}\right],(0.3,0.2,0.1)\\ \times\left[\frac{(0.5,0.2,0.2)-(0.5,0.2,0.1)}{(0.5,0.2,0.1)-(0.3,0.1,0.1)}\right]=(0.3,0.2,0.1),(0.3,0.2,0.2)\end{matrix}\right\}=(0.3,0.2,0.1)$$

$$F_2=\max\left\{\begin{matrix}(0.3,0.2,0.1)\times\left[\frac{(0.5,0.2,0.1)-(0.5,0.2,0.1)}{(0.5,0.2,0.1)-(0.3,0.2,0.1)}\right],(0.3,0.2,0.1)\\ \times\left[\frac{(0.5,0.2,0.2)-(0.4,0.2,0.1)}{(0.5,0.2,0.1)-(0.3,0.1,0.1)}\right]=(0.3,0.2,0.1),(0.2,0.2,0.2)\end{matrix}\right\}=(0.3,0.2,0.1)$$

$$F_3=\max\left\{\begin{matrix}(0.3,0.2,0.1)\times\left[\frac{(0.5,0.2,0.1)-(0.3,0.2,0.1)}{(0.5,0.2,0.1)-(0.3,0.2,0.1)}\right],(0.3,0.2,0.1)\\ \times\left[\frac{(0.5,0.2,0.2)-(0.3,0.1,0.2)}{(0.5,0.2,0.1)-(0.3,0.1,0.1)}\right]=(0.3,0.3,0.2),(0.3,0.2,0.2)\end{matrix}\right\}=(0.3,0.2,0.2)$$

$E_{i_{\max}}$, E_F and $F_{i_{\max}}$ are
$E_{i_{\min}}$ =(0.3,0.2,0.2), $E_{i_{\max}}$ =(0.3,0.2,0.1)
$F_{i_{\max}}=(0.3,0.2,0.1)$, $F_{i_{\min}}=(0.3,0.2,0.2)$

Step 3: Calculate the value of P_i

$$P_1 = (0.3,0.2,0.1) \times \left[\frac{(0.3,0.2,0.1)-(0.3,0.2,0.2)}{((0.3,0.2,0.1))-(0.3,0.2,0.2)} \right] + (0.7,0.8,0.9)$$

$$\times \left[\frac{(0.3,0.2,0.1)-(0.3,0.2,0.1)}{(0.3,0.2,0.1)-(0.3,0.2,0.2)} \right] = (0.3,0.2,0.2)+(0.7,0.8,0.9) = (0.7,0.2,0.2)$$

$$P_2 = (0.3,0.2,0.1) \times \left[\frac{(0.3,0.2,0.1)-(0.3,0.2,0.2)}{((0.3,0.2,0.1))-(0.3,0.2,0.2)} \right] + (0.7,0.8,0.9)$$

$$\times \left[\frac{(0.3,0.2,0.1)-(0.3,0.2,0.1)}{(0.3,0.2,0.1)-(0.3,0.2,0.2)} \right] = (0.3,0.2,0.2)+(0.7,0.8,0.9) = (0.7,0.2,0.2)$$

$$P_3 = (0.3,0.2,0.1) \times \left[\frac{(0.3,0.2,0.2)-(0.3,0.2,0.2)}{((0.3,0.2,0.1))-(0.3,0.2,0.2)} \right] + (0.7,0.8,0.9)$$

$$\times \left[\frac{(0.3,0.2,0.2)-(0.3,0.2,0.1)}{(0.3,0.2,0.1)-(0.3,0.2,0.2)} \right] = (0.3,0.2,0.1)+(0.7,0.8,0.9) = (0.7,0.2,0.1)$$

Step 4: Arrange the alternatives in ascending order

$$P_1 = P_2 < P_3$$

Step 5: The final ranking of the alternatives

$$A_3 > A_1 = A_2$$

Therefore, the third lathe model (A_3) is the best alternative.

7.6 CONCLUSION

The role of PFSM is highly significant. It can be applied to various MCDM problems and several situations involving more than two answers. We plan to extend this idea to rough set theory with two approximations and to provide a comparative analysis that reveals which picture fuzzy matrix either in soft setting or rough setting will be a more effective tool to deal with decision-making situations.

REFERENCES

Ashraf, S., S. Abdullah and A. Qadir. 2018a. Novel concept of cubic picture fuzzy sets. *Journal of New Theory* 24: 59–72.

Ashraf, S., T. Mahmood, S. Abdullah and Q. Khan. 2018b. Picture fuzzy linguistic sets and their applications for multi-attribute group decision making problems. *The Nucleus* 55, no.2: 66–73.

Atanassov, K.T. 1986. Intuitionistic fuzzy sets. *Fuzzy Sets and Systems* 20, no.1: 87–96.

Cagman, N. and S. Enginoglu. 2010. Soft matrix theory and its decision making. *Computers & Mathematics with Applications* 59, no. 10: 3308–3314.

Cuong, B.C. 2014. Picture fuzzy sets. *Journal of Computer Science and Cybernetics*. 30, no. 4: 409–420.

Dutta, P. and S. Ganju. 2018. Some aspects of picture fuzzy set. *Transactions of A. Razmadze Mathematical Institute* 172: 164–175.

Feng, Q. and Y. Zhou. 2014. Soft discernibility matrix and its application in decision making. *Applied Soft Computing* 24: 749–756.

Ganie, A.H., S. Singh and P. Bhatia. 2020. Some new correlation coefficients of picture fuzzy sets with applications. *Neural Computing and Applications* 32: 12609–12625.

Khan, M., P. Kumam, P. Liu, W. Kumam and U.R. Habib. 2020. An adjustable weighted soft discernibility matrix based on generalized picture fuzzy soft set and its applications in decision making. *Journal of Intelligent and Fuzzy Systems* 38, no. 2: 2103–2118.

Molodtsov, D. 1999. Soft set theory-first result. *Computers & Mathematics with Applications* 37: 19–31.

Opricovic, S. 1998. *Multicriteria Optimization of Civil Engineering Systems*. Faculty of Civil Engineering, Belgrade.

Singh, P. 2014. Correlation coefficients for picture fuzzy sets. *Journal of Intelligent and Fuzzy Systems* 27: 2857–2868.

Thong, N.T. 2015. HIFCF: an effective hybrid model between picture fuzzy clustering and intuitionistic fuzzy recommender systems for medical diagnosis expert systems with applications. *Expert System with Applications* 42: 3682–3701.

Wei, G.W. 2016. Picture fuzzy cross-entropy for multiple attribute decision making problems. *Journal of Business Economics and Management* 17: 491–502.

Wei, G.W. 2017. Some cosine similarity measures for picture fuzzy sets and their applications to strategic decision making. *Informatica* 28, no. 3: 547–564.

Yang, Y., C. Liang, S. Ji and T. Liu. 2015. Adjustable soft discernibility matrix based on picture fuzzy soft sets and its applications in decision making. *Journal of Intelligent and Fuzzy Systems* 29, no.4: 1711–1722.

Zadeh, L.A. 1965. Fuzzy sets. *Information and Control* 8, no. 3: 338–353.

8 Cubic n-Inner Product Space

S. Vijayabalaji
University College of Engineering Panruti
(A Constituent College of Anna University)

S. Sivaramakrishnan
Manakula Vinayagar Institute of Technology

P. Balaji
MEASI Academy of Architecture

CONTENTS

8.1 INTRODUCTION

Functional analysis is one of the significant branches in Mathematics with interesting structures such as normed linear space, inner product space and metric space. (Gähler 1965) made a thorough study of these structures and gave remarkable development in n-dimension version namely n-normed linear space (*n-nls*) and n-inner product space (*n- ips*). Zadeh (1965) in his inspiring work introduced the notion of fuzzy set theory. This motivated several researchers to focus their mind toward the development of various structures from the theory of functional analysis to fuzzy setting. In the papers (Bag and Samanta 2005, Chang and Mordesen 1994, Felbin 1992, Cho, Lin and Kim 2001, Dimmine, Gähler and White 1973, Katsaras 1984, Krishna and Sharma 1994, Kohli and Kumar 1993, Ming 1985), origin and creation of fuzzy inner product space and fuzzy normed linear space can be seen. The *n-nls* and *n-ips* were generalized to fuzzy n-normed linear space (*f-n-nls*) and fuzzy-n-inner product space (*f-n-ips*) by (Vijayabalaji and Thillaigovindan 2007a). They studied about several interesting properties of *f-n-nls* and *f-n-ips* and their inter-relationship in several papers (Ming 1985, Misiak 1989, Narayanan and Vijayabalaji 2005, Thillaigovindan, Vijayabalaji and Anita Shanthi 2010). Zadeh (1975) also generalized the idea of fuzzy sets to interval-valued fuzzy sets. Motivated by this theory, Vijayabalaji

(2011) developed the idea of equivalent fuzzy n-inner product space. Cubic set is a pioneering structure that was introduced and developed by (Jun et al. 2010, 2011). The novelty of this structure is that it comprises of fuzzy set and interval-valued fuzzy set. Several algebraic and analytic structure emerged using this theory. Cubic n-normed linear space (C-*n-nls*) is one among the interesting concepts that were developed by Vijayabalaji (2017). *C-n-nls* is the structure that comprises *f-n-nls* and *iv-f-nls*. Vijayabalaji and Sivaramakrishnan (2015) introduced the notion of cubic linear space. The above theories inspired for the development of the present topic namely cubic n-inner product space (*C-n-ips*). *C-n-ips* is an interesting structure that comprises fuzzy n-inner product space (*f-n-ips*) and interval-valued fuzzy n-inner product space (*i-v-f-n-ips*). We formulate a new structure called a cubic intuitionistic linear space and define some operations followed by examples. We form the remaining sections as preliminaries, *i-v-f-n-ips, C-n-ips* and cubic n-inner product space.

8.2 PRELIMINARIES

For convenience, let us have the following notations in the entire paper while denoting the elements in an n-normed linear space $\left(X,\|\bullet,\ldots,\bullet\|\right)$. Normally for a given linear space X, the elements in X^n are denoted by $\left(x_1,x_2,\ldots,x_{n-1},x_n\right)$ (or) $\left(x_1,x_2,\ldots,x_{n-1},x_n\right)$. Let us denote this by $\left(\boldsymbol{H}_{n-1},x_n\right)$ (or) $\left(\boldsymbol{H}_n\right)$ wherever applicable.

Likewise, consider the n-inner product space $(X,(\bullet,\bullet \mid \bullet,\ldots,\bullet))$. The elements namely $(x,x \mid x_2,\ldots,x_n)$ (or) $(x,y \mid x_2,\ldots,x_n)$ are denoted by $\left(\boldsymbol{L}_{xx}\right)$ (or) $\left(\boldsymbol{L}_{xy}\right)$.

Definition 8.1

For a given linear space X, a function $\|\bullet,\ldots,\bullet\|$ defined on X^n is called n-norm if it satisfies the following four conditions.

1. $\|H_n\| = 0$ if and only if all the n elements are linearly dependent.
2. $\|\boldsymbol{H}_n\|$ is invariant under any permutation of n elements.
3. $\left\|\boldsymbol{H}_{n-1},cx_n\right\| = |c|\left\|H_n\right\|$, for any real scalar c.
4. $\left\|\boldsymbol{H}_{n-1},y+z\right\| \le \left\|H_{n-1},y\right\| + \left\|H_{n-1},z\right\|$

$\left(X,\|\bullet,,\ldots,\bullet,\|\right)$ will be termed as n-normed linear space.

Definition 8.2

A mapping of M from **X** into the unit interval [0,1] is called a fuzzy set of **X.**

Definition 8.3

For a given nonempty set $\boldsymbol{X}$, a cubic set is represented by $\left(x,\overline{B},B\right)$ where $x \in \boldsymbol{X}$, $\overline{B}$ is an interval-valued fuzzy set of $\boldsymbol{X}$ and B is fuzzy set on $\boldsymbol{X}$.

Definition 8.4

A fuzzy set N from $X^n \times R \rightarrow [0,1]$ is called a fuzzy n-normed linear space (X, N) with $t \in R$, and the following conditions are true.

1. $N(\boldsymbol{H}_n, t) = 0$, where $t \leq 0$
2. $N(\boldsymbol{H}_n, t) = 1$ if all the n elements are linearly dependent. (Here $t > 0$)
3. $N(\boldsymbol{H}_n, t)$ is invariant with respect to the permutation of n elements.
4. $N(\boldsymbol{H}_{n-1}, cx_n, t) = N\left(\boldsymbol{H}_{n-1}, x_n, \frac{t}{|c|}\right)$, if $c \neq 0, c \in F$ (field), $t > 0$.
5. For all $s, t \in R$,

$$N\left(\boldsymbol{H}_{n-1}, x_n + x_n', s + t\right) \geq \min\left\{N(\boldsymbol{H}_{n-1}, x_n, s), N\left(\boldsymbol{H}_{n-1}, x_n', t\right)\right\}$$

6. $N(H_n, t)$ is an increasing function of $t \in R$ and $\lim_{t \to \infty} N(\boldsymbol{H}_n, t) = 1$.

Re-structure of the above definition using t-norm is as follows.

Definition 8.5

A fuzzy n – norm on X is a fuzzy set N from $X^n \times [0, \infty) \rightarrow [0,1]$ subject to

1. $N(\boldsymbol{H}_n, t) > 0$.
2. $N(\boldsymbol{H}_n, t) = 1$ if all the n elements are linearly dependent.
3. $N(\boldsymbol{H}_n, t)$ is invariant under any permutation of n elements.
4. $N(\boldsymbol{H}_{n-1}, cx_n, t) = N\left(\boldsymbol{H}_{n-1}, x_n, \frac{t}{|c|}\right)$, if $c \neq 0, c \in F$ (field).
5. $N\left(\boldsymbol{H}_{n-1}, x_n + x_n', s + t\right) \geq \left\{N(\boldsymbol{H}_{n-1}, x_n, s) * N\left(\boldsymbol{H}_{n-1}, x_n', t\right)\right\}$.
6. $N(\boldsymbol{H}_n, t)$ is left continuous and nondecreasing function such that

$$\lim_{t \to \infty} N(\boldsymbol{H}_n, t) = 1$$

(X, N) is called fuzzy n-normed linear space.

Definition 8.6

A mapping from a nonempty set to the set of all intervals is called an interval-valued fuzzy set.

Definition 8.7

For a given linear space X with the two functions $N: X^n \times [0, \infty) \rightarrow [0,1]$ and $\bar{N}: X^n \times [0, \infty) \rightarrow \boldsymbol{D}[0,1]$, the fuzzy set and the interval-value fuzzy set are referred

to as cubic n-normed linear space, respectively, or briefly cubic n-NLS and represented by
$C = (X, N, \bar{N})$ if it possesses the following conditions.

1. $N(\boldsymbol{H}_n, t) > 0$.
2. $N(\boldsymbol{H}_n, t) = 0$ if all the n elements are linearly dependent.
3. $N(\boldsymbol{H}_n, t)$ is invariant under any permutation of n elements.
4. $N(\boldsymbol{H}_{n-1}, cx_n, t) = N\left(\boldsymbol{H}_{n-1}, x_n, \frac{t}{|c|}\right)$, if $c \neq 0, c \in F$ (field)
5. $N(\boldsymbol{H}_{n-1}, x_n + x_n', s + t) \leq N(\boldsymbol{H}_{n-1}, x_n, s) N(\boldsymbol{H}_{n-1}, x_n', t)$.
6. $N(\boldsymbol{H}_n, t)$ is left continuous and nonincreasing function of $t \in R$ such that

$$\lim_{t \to \infty} N(\boldsymbol{H}_n, t) = 0$$

7. $\bar{N}(\boldsymbol{H}_n, t) > \bar{0}$.
8. $\bar{N}(\boldsymbol{H}_n, t) = \bar{1}$ if all the n are linearly dependent.
9. $\bar{N}(\boldsymbol{H}_n, t))$ is invariant under any permutation of n elements.
10. $\bar{N}(\boldsymbol{H}_{n-1}, cx_n, t) = \bar{N}\left(\boldsymbol{H}_{n-1}, x_n, \frac{t}{|c|}\right)$, if $c \neq 0, c \in F$(field).
11. $\bar{N}(\boldsymbol{H}_{n-1}, x_n + x_n', s + t) \geq \bar{N}(\boldsymbol{H}_{n-1}, x_n, s) * \bar{N}(\boldsymbol{H}_{n-1}, x_n', t)$.
12. $\bar{N}(\boldsymbol{H}_n, t)$ is left continuous and nondecreasing function of $t \in R$ such that

$$\lim_{t \to \infty} \bar{N}(\boldsymbol{H}_n, t) = \bar{1}$$

Definition 8.8

Given a linear space X and a real valued function $(\bullet, \bullet \mid \bullet, \ldots, \bullet)$ on X^{n+1}, we define n-inner product space $(X, (\bullet, \bullet \mid \bullet, \ldots, \bullet))$ with the following conditions.

1. (i)$(\boldsymbol{L}_{xx}) \geq 0$,
   (ii)$(\boldsymbol{L}_{xx}) = 0$ iff all the $n+1$ elements are linearly dependent.
2. $(\boldsymbol{L}_{xy}) = (\boldsymbol{L}_{yx})$,
3. $(\boldsymbol{L}_{xy})$ is invariant under any permutation of $n-1$ elements,
4. $(\boldsymbol{L}_{xx}) = (\boldsymbol{L}_{x_2 x_2})$,
5. $(\boldsymbol{L}_{ax}) = a(\boldsymbol{L}_x)$ for every $a \in R$ (real),
6. $(\boldsymbol{L}_{x+x'}) = (\boldsymbol{L}_x) + (\boldsymbol{L}_{x'})$.

Definition 8.9

A fuzzy set $S : X^{n+1} \times R$ (set of real numbers) is called a fuzzy n-inner product space (X, S) for a given linear space X with $t \in R$, if it holds the below conditions.

1. $S(L_{xx},t)=0$ ($t\le 0$).
2. $S(L_{xx},t)=1$ if $n+1$ elements are linearly dependent ($t>0$).
3. $S(L_{xy},t))=S(L_{yx},t)$ ($t>0$).
4. $S(L_{xy},t)$ is invariant under any permutation of $n-1$ elements.
5. $S(L_{xx},t)=S(L_{x2,x2},t)$($t>0$).
6. $S(L_{ax,bx},t)=J\left(L_{x,x},\dfrac{t}{|ab|}\right)$, $a,b\in R$, $t>0$
7. For all $s,t\in R$,

$$S(L_{x+x'},t+s)\ge \min\{S(L_x,t),S(L_{x'},s)\}$$

8. For all $s,t\in R$ with $s>0$, $t>0$,

$$S(L_{xy},\sqrt{ts})\ge \min\{S(L_x,t),S(L_y,s)\}$$

9. $S(L_{xy},t)$ is a nondecreasing function of $t\in R$ and $\lim_{t\to\infty} S(L_{xy},t)=1$.

Definition 8.10

Let X be a linear space over a field F, $(X,\bar{\rho})$ be an interval-valued fuzzy linear space, and (X,ε) be a fuzzy linear space. A cubic set $C=\langle\bar{\rho},\varepsilon\rangle$ in X is called a cubic linear space of X if it satisfies for all $j,k\in X$ and $c_1,c_2\in F$ (field).

i. $\bar{\rho}(c_1 j\odot c_2 k)\ge \min\{\bar{\rho}(j),\bar{\rho}(k)\}$,
ii. $\varepsilon(c_1 j\odot c_2 k)\le \max\{\varepsilon(j),\varepsilon(k)\}$.

8.3 CUBIC n-INNER PRODUCT SPACE

Definition 8.11

By a cubic n-inner product space (*C-n-ips*), we mean a structure of the form = $(X,S,\bar{S})$, where $S: X^{n+1}\times[0,\infty)\to[0,1]$ is a fuzzy set and $\bar{S}: X^{n+1}\times[0,\infty)\to \boldsymbol{D}[0,1]$ is an interval-valued fuzzy set satisfying the following properties.

1. $S(L_{xx},t)>0$.
2. $S(L_{xx},t)=0$ if $n+1$ is linearly dependent.
3. $S(L_{xy},t)=S(L_{yx},t)$.
4. $S(L_{xy},t)$ is invariant under any permutation of $n-1$ elements.
5. $S(L_{xx},t)=S(L_{x2x2},t)$.
6. For all $t>0$, $S(L_{axbx},t)=S\left(L_{x,x},\dfrac{t}{|ab|}\right)$, $a,b\in R$ (real)
7. $S(L_{(x+x')y},t+s)\le S(L_{xy},t)S(L_{x'y},s)$

8. $S\left(L_{xy},\sqrt{ts}\right) \le S(L_{xx},t)S\left(L_{yy},s\right)$
9. $S(L_{xy} \mid x_2,\ldots,x_n,t)$ is a nonincreasing function of $t \in R$ and $\lim_{t\to\infty} S\left(L_{xy},t\right)=0$.
10. $\bar{S}(L_{xx},t)>0$.
11. $\bar{S}(L_{xx},t)=0$ if $n+1$ elements are linearly dependent.
12. $\bar{S}\left(L_{xy},t\right)=\bar{S}\left(L_{yx},t\right)$.
13. $\bar{S}\left(L_{xy},t\right)$ is invariant under any permutation of $n-1$ elements.
14. $\bar{S}(L_{xx},t)=\bar{S}\left(L_{x_2x_2},t\right)$.
15. $\bar{S}(L_{axbx},t)=\bar{S}\left(L_{xx},\frac{t}{|ab|}\right)$, $a,b \in R, t>0$
16. $\bar{S}\left(L_{(x+x')y},t+s\right) \ge \bar{S}\ (L_{xy},t) * \bar{S}\left(L_{x'y},s\right)$
17. $\bar{S}\left(L_{xy},\sqrt{ts}\right) \ge \bar{S}(L_{xx},t) * \bar{S}\left(L_{yy},s\right)$
18. $\bar{S}\left(L_{xy},t\right)$ is an increasing function of $t \in R$ and $\lim_{t\to\infty} \bar{S}\left(L_{xy},t\right)=0$.

Example 8.1

Let $(X,(\bullet,\bullet \mid \bullet,\ldots,\bullet))$ be an n-inner product space. Define $\boldsymbol{a} * \boldsymbol{b} = \mathbf{min}\{\boldsymbol{a},\boldsymbol{b}\}$, $\boldsymbol{a} \Diamond \boldsymbol{b} = \mathbf{max}\{\boldsymbol{a},\boldsymbol{b}\}$ for $\boldsymbol{a},\boldsymbol{b} \in [0,1]$, $S\left(L_{xy},t\right)=\frac{\left(L_{xy}\right)}{t+\left(L_{xy}\right)}$ and $\bar{S}\left(L_{xy},t\right)=\frac{t}{t+\left(L_{xy}\right)}$. Then $C=\left(X,I,\bar{I}\right)$ is a *C-n-ips*.

Theorem 8.1

The *R*-intersection, $(C_1 \cap C_2)_R = \left(X,\bar{S}\cap\bar{S},S_1 \cup S_2\right)$ of two *C-n-ips* is again a *C-n-ips*.

Proof: Define $\left(S_1 \cup S_2\right)\left(L_{xy},t\right)=\max\left\{S_1\left(L_{xy},t\right),S_2\left(L_{xy},t\right)\right\}$ and

$$\left(\bar{S}\cap\bar{S}\right)\left(L_{xy},t\right)=\min\left\{\bar{S}\left(L_{xy},t\right),\bar{S}\left(L_{xy},t\right)\right\}$$

To validate this, we proceed as follows.

1. As $S_1\left(L_{xy},t\right)>0$ and $S_2\left(L_{xy},t\right)>0$, it follows that $\left(S_1 \cup S_2\right)\left(L_{xy},t\right)>0$.
2. $\left(S_1 \cup S_2\right)\left(L_{xy},t\right)=0$

$$\Leftrightarrow \max\{S_1\left(L_{xy},t\right)=0,S_2\left(L_{xy},t\right)\}=0$$

$$\Leftrightarrow S_1\left(L_{xy},t\right)=0,S_2\left(L_{xy},t\right)=0$$

$\Leftrightarrow$ n+1 elements are linearly dependent.

3. $S_1(L_{xy},t)$ and $S_2(L_{xy},t)$ are invariant under any permutation of $n-1$ elements, it follows that $(S_1 \bigcup S_2)(L_{xy},t)$ is invariant under any permutation of $n-1$ elements.

   (4) and (5) follows directly from the definition.

6. $(S_1 \bigcup S_2)(L_{axby},t) = \max\{S_1(L_{axby},t),\ S_2(L_{L_{axby}},t)\}$

$$= \max\left\{S_1\left(L_{x,y},\frac{t}{|ab|}\right), S_2\left(L_{x,y},\frac{t}{|ab|}\right)\right\}$$

$$= (S_1 \bigcup S_2)\left(L_{x,y},\frac{t}{|ab|}\right)$$

7. $(S_1 \bigcup S_2)((_{(x+x')y},\ t + s)$

$$= \max\left\{S_1\left((_{(x+x')y},\ t + s\right), S_2(L_{(x+x')y},\ t + s)\right\}$$

$$\leq \max\left\{\max\left[S_1(L_{xy},t), S_1(L_{x'y},s)\right], \max\left[S_2(L_{xy},t), S_2(L_{x'y},s)\right]\right\}$$

$$= \max\left\{\max\left[S_1(L_{xy},t), S_2(L_{x'y},t)\right], \max\left[S_1(L_{xy},s), S_2(L_{x'y},s)\right]\right\}$$

$$= \max\left\{\left(S_1 \bigcup S_2\right)(L_{x,y},t), (S_1 \bigcup S_2)\ (L_{x'y},s)\right\}$$

Thus,

$$\left(S_1 \bigcup S_2\right)(L_{(x+x')y},\ t + s)$$

$$\leq \max\left\{\left(S_1 \bigcup S_2\right)(L_{x,y},t), (S_1 \bigcup S_2)\ (L_{x'y},s)\right\}$$

Similarly, we can verify all the conditions of the *C-n-ips*.

Remark 8.1

i. Let $C_1 = \left(X, S_1, \overline{S_1}\right)$ and $C_2 = \left(X, S_2, \overline{S_2}\right)$ be two cubic n-inner product spaces. Then their *R*-union $(C_1 \bigcup C_2)_R = \left(X, \overline{S} \bigcup \overline{S}, S_1 \bigcap S_2\right)$ need not be a *C-n-ips*.

ii. Let $C_1 = \left(X, S_1, \overline{S_1}\right)$ and $C_2 = \left(X, S_2, \overline{S_2}\right)$ be two cubic n-inner product spaces. Then their P-intersection $(C_1 \bigcap C_2)_P = \left(X, \overline{S} \bigcap \overline{S}, S_1 \bigcap S_2\right)$ need not be a C-n-ips.

iii. Let $C_1 = \left(X, S_1, \overline{S_1}\right)$ and $C_2 = \left(X, S_2, \overline{S_2}\right)$ be two cubic n-inner product spaces. Then their P-union $(C_1 \bigcup C_2)_P = \left(X, \overline{S} \bigcup \overline{S}, S_1 \bigcup S_2\right)$ need not be a C-n-ips.

8.4 CUBIC INTUITIONISTIC LINEAR SPACE

Definition 8.12

Let X be a linear space with a binary operation $\odot$ over a field F. Let $\overline{\omega} = \langle V, \overline{\rho}, \overline{\sigma} \rangle$ be an interval-valued intuitionistic fuzzy linear space, and let $\omega = \langle X, \varepsilon, \zeta \rangle$ be an intuitionistic fuzzy linear space. A cubic intuitionistic set $C = \langle \overline{\omega}, \omega \rangle$ in X is called a cubic intuitionistic linear space of X if for all $j, k \in X$ and $c_1, c_2 \in F$,

i. $\overline{\rho}\left(c_1\ j \odot c_2\ k\right) \geq \min\left\{\overline{\rho}(j), \overline{\rho}(k)\right\}$,
ii. $\overline{\sigma}\left(c_1\ j \odot c_2\ k\right) \leq \max\left\{\overline{\sigma}(j), \overline{\sigma}(k)\right\}$,
iii. $\varepsilon\left(c_1\ j \odot c_2\ k\right) \leq \max\left\{\varepsilon(j), \varepsilon(k)\right\}$,
iv. $\zeta\left(c_1\ j \odot c_2\ k\right) \geq \min\left\{\zeta(j), \zeta(k)\right\}$.

Example 8.2

Let the Klein 4-group defined by the binary operation $\odot$ be $X = \{i, j, k, m\}$ as follows:

Let F be the field GF (2). Let $(0)h = i$, $(1)h = w$ for all $w \in X$. Then X is a linear space over F.

Define an interval-valued intuitionistic fuzzy set $\overline{\omega} = \overline{\rho}, \overline{\sigma}$ in X by $\overline{\rho}(e) = [0.8, 0.9]$,

$$\overline{\rho}(j) = [0.4, 0.5] = \overline{\rho}(k), \overline{\rho}(z) = [0.6, 0.8] \text{and}$$

$$\overline{\sigma}(x) = \begin{cases} [0.3,\ 0.4],\ \textit{if}\ x = i \\ [0.9,\ 1],\ \text{otherwise.} \end{cases}$$

TABLE 8.1
Cayley Table

$\odot$	i	j	k	m
i	i	j	k	m
j	j	i	m	k
k	k	m	i	j
m	m	k	j	i

Then $\bar{\omega} = \langle \bar{\rho}, \bar{\sigma} \rangle$ is an interval-valued intuitionistic fuzzy linear space.

Define an intuitionistic fuzzy set $\omega = \langle \varepsilon, \zeta \rangle$ in X by

$$\varepsilon(x) = \begin{cases} 0.4, & \textit{if } x = i \\ 0.9, & \text{otherwise} \end{cases}$$

and

$$\zeta(x) = \begin{cases} 0.9, & \textit{if } x = i \\ 0.5, & \text{otherwise} \end{cases}$$

Note that $\omega = \langle \varepsilon, \zeta \rangle$ is an intuitionistic fuzzy linear space of X.

Therefore, $C = \langle \bar{\omega}, \omega \rangle$ is a cubic intuitionistic linear space of X.

Theorem 8.2

Let $C_1 = (\bar{\omega}_1, \omega_1)$ and $C_2 = (\bar{\omega}_2, \omega_2)$ be two cubic intuitionistic linear spaces, where $\bar{\omega}_1 = \langle \bar{\rho}_1, \bar{\sigma}_1 \rangle$, $\bar{\omega}_2 = \langle \bar{\rho}_2, \bar{\sigma}_2 \rangle$, $\omega_1 = \langle \varepsilon_1, \zeta_1 \rangle$ and $\omega_2 = \langle \varepsilon_2, \zeta_2 \rangle$. Then their R-intersection, $\left(C_1 \bigcap C_2\right)_R = \left(\bar{\omega}_1 \bigcap \bar{\omega}_2, \omega_1 \bigcup \omega_2\right)$ is a cubic intuitionistic linear space.

Proof: Define $\bar{\omega}_1 \bigcap \bar{\omega}_2$ as

$$\left(\bar{\omega}_1 \bigcap \bar{\omega}_2\right)(c_1\ j \odot c_2\ k) = \left\{\left(\bar{\rho}_1 \bigcap \bar{\rho}_2\right)(c_1\ j \odot c_2\ k), \left(\bar{\sigma}_1 \bigcup \bar{\sigma}_2\right)(c_1\ j \odot c_2\ k)\right\}$$

and

Define $\omega_1 \bigcup \omega_2$ as

$$\left(\omega_1 \bigcup \omega_2\right)(c_1 j \odot c_2\ k) = \left\{\left(\varepsilon_1 \bigcup \varepsilon_2\right)(c_1\ j \odot c_2\ k), \left(\zeta_1 \bigcap \zeta_2\right)(c_1\ j \odot c_2\ k)\right\}$$

Then

i. $\left(\bar{\rho}_1 \bigcap \bar{\rho}_2\right)(c_1\ j \odot c_2\ k) = \min\ \left\{\bar{\rho}_1(c_1\ j \odot c_2\ k),\ \bar{\rho}_2(c_1\ j \odot c_2\ k)\right\}$

$$\geq \min\left\{\min\left[\bar{\rho}_1(j), \bar{\rho}_1(k)\right], \min[\bar{\rho}_2(j), \bar{\rho}_2(k)]\right\}$$

$$= \min\ \left\{\min\left[\bar{\rho}_1(j),\ \bar{\rho}_2(j)\right], \min\left[\bar{\rho}_1(k), \bar{\rho}_2(k)\right]\right\}$$

$$= \min\left\{\left(\bar{\rho}_1 \bigcap \bar{\rho}_2\right)(j), \left(\bar{\rho}_1 \bigcap \bar{\rho}_2\right)(k)\right\}$$

$$\Rightarrow \left(\bar{\rho}_1 \bigcap \bar{\rho}_2\right)(c_1\ j \odot c_2\ k) \geq \min\left\{\left(\bar{\rho}_1 \bigcap \bar{\rho}_2\right)(j), \left(\bar{\rho}_1 \bigcap \bar{\rho}_2\right)(k)\right\}$$

ii. $\left(\bar{\sigma}_1 \bigcup \bar{\sigma}_2\right)(c_1\ j \odot c_2\ k) = \max\left\{\bar{\sigma}_1(c_1\ j \odot c_2\ k), \bar{\sigma}_2(c_1\ j \odot c_2\ k)\right\}$

$$\leq \max\left\{\max\left[\bar{\sigma}_1(j), \bar{\sigma}_1(k)\right], \max\left[\bar{\sigma}_2(j), \bar{\sigma}_2(k)\right]\right\}$$

$$= \max\left\{\max\left[\bar{\sigma}_1(j), \bar{\sigma}_2(j)\right], \max\left[\bar{\sigma}_1(k), \bar{\sigma}_2(k)\right]\right\}$$

$$= \max\left\{\left(\bar{\sigma}_1 \bigcup \bar{\sigma}_2\right)(j), \left(\bar{\sigma}_1 \bigcup \bar{\sigma}_2\right)(k)\right\}$$

$$\Rightarrow \left(\bar{\sigma}_1 \bigcup \bar{\sigma}_2\right)(c_1\ j \odot c_2\ k) \leq \max\left\{\left(\bar{\sigma}_1 \bigcup \bar{\sigma}_2\right)(j), \left(\bar{\sigma}_1 \bigcup \bar{\sigma}_2\right)(k)\right\}$$

iii. $\left(\varepsilon_1 \bigcup \varepsilon_2\right)(c_1\ j \odot c_2\ k) = \max\left\{\varepsilon_1(c_1\ j \odot \beta k), \varepsilon_2(c_1\ j \odot c_2\ k)\right\}$

$$\leq \max\left\{\max\left[\varepsilon_1(j), \varepsilon_1(k)\right], \max\left[\varepsilon_2(j), \varepsilon_2(k)\right]\right\}$$

$$= \max\left\{\max\left[\varepsilon_1(j), \varepsilon_2(j)\right], \max\left[\varepsilon_1(k), \varepsilon_2(k)\right]\right\}$$

$$= \max\left\{\left(\varepsilon_1 \bigcup \varepsilon_2\right)(j), \left(\varepsilon_1 \bigcup \varepsilon_2\right)(k)\right\}$$

$$\Rightarrow \left(\varepsilon_1 \bigcup \varepsilon_2\right)(c_1\ j \odot c_2\ k) \leq \max\left\{\left(\varepsilon_1 \bigcup \varepsilon_2\right)(j), \left(\varepsilon_1 \bigcup \varepsilon_2\right)(k)\right\}$$

iv. $\left(\zeta_1 \bigcap \zeta_2\right)(c_1\ j \odot c_2\ k) = \min\left\{\zeta_1(c_1\ j \odot \beta k), \zeta_2(c_1\ j \odot c_2\ k)\right\}$

$$\geq \min\left\{\min\left[\zeta_1(j), \zeta_1(k)\right], \min\left[\zeta_2(j), \zeta_2(k)\right]\right\}$$

$$= \min\left\{\min\left[\zeta_1(j),)\zeta_2(j)\right], \min\left[\zeta_1(k), \zeta_2(k)\right]\right\}$$

$$= \min\left\{(\zeta_1 \bigcap \zeta_2(j)\left(\zeta_1 \bigcap \zeta_2\right)(k)\right\}$$

$$\Rightarrow \left(\zeta_1 \bigcap \zeta_2\right)(c_1\ j \odot c_2\ k) \geq \min\left\{(\zeta_1 \bigcap \zeta_2)(j)\left(\zeta_1 \bigcap \zeta_2\right)(k)\right\}$$

Thus, $\left(C_1 \bigcap C_2\right)_R = \left(\bar{\omega}_1 \bigcap \bar{\omega}_2, \omega_1 \bigcup \omega_2\right)$ is a cubic intuitionistic linear space .

Remark 8.2

i. Let $C_1 = (\bar{\omega}_1, \omega_1)$ and $C_2 = (\bar{\omega}_2, \omega_2)$ be two cubic linear spaces. Then their R-union
$(C_1 \bigcup C_2)_R = \left(\bar{\omega}_1 \bigcup \bar{\omega}_2, \omega_1 \bigcap \omega_2\right)$ need not be a cubic intuitionistic linear space.

ii. Let $C_1 = (\bar{\omega}_1, \omega_1)$ and $C_2 = (\bar{\omega}_2, \omega_2)$ be two cubic linear spaces. Then their P-intersection
$(C_1 \bigcap C_2)_P = \left(\bar{\omega}_1 \bigcap \bar{\omega}_2, \omega_1 \bigcap \omega_2\right)$ need not be a cubic intuitionistic linear space.

iii. Let $C_1 = (\bar{\omega}_1, \omega_1)$ and $C_2 = (\bar{\omega}_2, \omega_2)$ be two cubic linear spaces. Then their P-union
$(C_1 \bigcup C_2)_P = \left(\bar{\omega}_1 \bigcup \bar{\omega}_2, \omega_1 \bigcup \omega_2\right)$ need not be a cubic intuitionistic linear space.

Proof: Let $X = \{i, j, k, m\}$ be the Klein 4-group as in Example 8.2. Choose F as Galois field with two elements subject to $(0)w = e$, $(1)w = w$ for all $w \in X$. X being linear space over F

We frame two ivif-sets as, $\bar{\omega}_1 = \langle \bar{\rho}_1, \bar{\sigma}_1 \rangle$, $\bar{\omega}_2 = \bar{\rho}_2, \bar{\sigma}_2$

$$\bar{\rho}_1(i) = [0.79,\ 0.9],$$

$$\bar{\rho}_1(j) = [0.3, 0.45] = \bar{\rho}_1(k)$$

$$\bar{\rho}_1(m) = [0.6.07],$$

$$\bar{\rho}_2(i) = [0.7.0.1],$$

$$\bar{\rho}_2(j) = [0.2,\ 0.3],$$

$$\bar{\rho}_2(k) = [0.5,\ 0.6] = \bar{\rho}_2(m),$$

$$\bar{\sigma}_1(i) = [0.3,\ 0.4],$$

$$\bar{\sigma}_1(j) = [0.9,\ 1] = \bar{\sigma}_1(k)$$

$$\bar{\sigma}_1(m) = [0.5,\ 0.6] \text{and}$$

$$\bar{\sigma}_2(i) = [0.2, 0.3],$$

$$\bar{\sigma}_2(j) = [0.7,\ 0.8]$$

$$\bar{\sigma}_2(k) = [0.6,\ 0.7] = \bar{\sigma}_2(m).$$

Observe that $\bar{\omega}_1 = \langle \bar{\rho}_1, \bar{\sigma}_1 \rangle$, $\bar{\omega}_2 = \langle \bar{\rho}_2, \bar{\sigma}_2 \rangle$ are interval-valued intuitionistic fuzzy linear spaces of X.

Define

$$\left(\bar{\omega}_1 \bigcup \bar{\omega}_2\right)(x) = \left\{(\bar{\rho}_1 \bigcup \bar{\rho}_2)(x)\right\} \text{for all } x \in X, \text{ where}$$

$$(\bar{\rho}_1 \bigcup \bar{\rho}_2)(x) = \max\{\bar{\rho}_1(x), \bar{\rho}_2(x)\} \text{ and}$$

$$(\bar{\sigma}_1 \bigcap \bar{\sigma}_2)(x) = \min\{\bar{\sigma}_1(x), \bar{\sigma}_2(x)\}$$

So,

$$(\bar{\rho}_1 \bigcup \bar{\rho}_2)(i) = [0.79, 1],$$

$$\left(\bar{\rho}_1 \bigcup \bar{\rho}_2\right)(j) = [0.3, 0.45],$$

$$\left(\bar{\rho}_1 \bigcup \bar{\rho}_2\right)(k) = [0.5, 0.65],$$

$$\left(\bar{\rho}_1 \bigcup \bar{\rho}_2\right)(m) = [0.6, 0.7] \text{ and}$$

$$\left(\bar{\sigma}_1 \bigcap \bar{\sigma}_2\right)(i) = [0.2, 0.3],$$

$$\left(\bar{\sigma}_1 \bigcap \bar{\sigma}_2\right)(j) = [0.7, 0.8],$$

$$\left(\bar{\sigma}_1 \bigcap \bar{\sigma}_2\right)(k) = [0.6, 0.7],$$

$$\left(\bar{\sigma}_1 \bigcap \bar{\sigma}_2\right)(m) = [0.5, 0.6].$$

Thus, $\bar{\omega}_1 \bigcup \bar{\omega}_2$ is an interval-valued intuitionistic fuzzy subset of X.

We have,

$$(\bar{\rho}_1 \bigcup \bar{\rho}_2)(k \odot m) \min\left\{(\bar{\rho}_1 \bigcup \bar{\rho}_2)(k)(\bar{\rho}_1 \bigcup \bar{\rho}_2)(m)\right\}$$

$$\Rightarrow (\bar{\rho}_1 \bigcup \bar{\rho}_2)(j) \geq \min\{[0.5, 0.65], [0.6, 0.7]\}$$

But $(\bar{\rho}_1 \bigcup \bar{\rho}_2)(j) = [0.3, 0.45] \geq [0.5, 0.65]$, which is absurd.

Moreover,

$$(\bar{\sigma}_1 \bigcap \bar{\sigma}_2)(k \odot m) \leq \max\left\{(\bar{\sigma}_1 \bigcap \bar{\sigma}_2)(k), (\bar{\sigma}_1 \bigcap \bar{\sigma}_2)(m)\right\}$$

$$\Rightarrow (\bar{\sigma}_1 \bigcap \bar{\sigma}_2)(j) \leq \max\{\, [0.6, 0.7], [0.5, 0.6] \,\} = [0.6, 0.7]$$

But $(\bar{\sigma}_1 \cap \bar{\sigma}_2)(j) = [0.7, 0.8] \leq [0.6, 0.7]$

This is also contradictory.

This demonstrates that the union of two interval-valued intuitionistic fuzzy linear spaces need not be an interval-valued intuitionistic fuzzy linear space.

Defining two intuitionist fuzzy sets $\omega_1 = \langle \varepsilon_1, \zeta_1 \rangle$ and $\omega_2 = \langle \varepsilon_2, \zeta_2 \rangle$ in X by

$$\varepsilon_1(i) = 0.4, \varepsilon_1(j) = 0.5, \varepsilon_1(k) = \varepsilon_1(m) = 0.91,$$

$$\varepsilon_2(i) = 0.3, \varepsilon_2(j) = \varepsilon_2(m) = 0.8, \varepsilon_2(k) = 0.7,$$

$$\zeta_1(i) = 1, \zeta_1(j) = \zeta_1(k) = 0.5, \zeta_1(\mathrm{m}) = 0.7 \text{ and}$$

$$\zeta_2(i) = 0.9, \zeta_2(j) = 0.4, \zeta_2(k) = \zeta_2(\mathrm{m}) = 0.6.$$

We observe that $\omega_1 = \langle \varepsilon_1, \zeta_1 \rangle>$ and $\omega_2 = \langle \varepsilon_2, \zeta_2 \rangle$ are intuitionistic fuzzy linear spaces of X.

Define

$$\left(\bar{\omega}_1 \cap \bar{\omega}_2\right)(x) = \left\{\left(\varepsilon_1 \cap \varepsilon_2\right)(x), \left(\zeta_1 \cup \zeta_2\right)(x)\right\} \text{all } x \in X, \text{Where,}$$

$$\left(\varepsilon_1 \cap \varepsilon_2\right)(x) = \min\{\varepsilon_1(x),\ \varepsilon_2(x)\} \text{and}$$

$$\left(\zeta_1 \cup \zeta_2\right)(x) = \max\{\zeta_1(x),\ \zeta_2(x)\}$$

Then

$$\left(\varepsilon_1 \cap \varepsilon_2\right)(i) = 0.3,$$

$$\left(\varepsilon_1 \cap \varepsilon_2\right)(j) = 0.5, \left(\varepsilon_1 \cap \varepsilon_2\right)(k) = 0.7,$$

$$\left(\varepsilon_1 \cap \varepsilon_2\right)(m) = 0.8 \text{ and}$$

$$\left(\zeta_1 \cup \zeta_2\right)(i) = 1,$$

$$\zeta_1 \cup \zeta_2(j) = 0.5 \text{ and } \left(\zeta_1 \cup \zeta_2\right)(k) = 0.6,$$

$$\left(\zeta_1 \cup \zeta_2\right)(m) = 0.7$$

So $\left(\omega_1 \bigcap \omega_2\right)$ is intuitionistic fuzzy subset of X.

$$\left(\varepsilon_1 \bigcap \varepsilon_2\right)(j \odot K) \le \max\left\{\left(\varepsilon_1 \bigcap \varepsilon_2\right)(j), \left(\varepsilon_1 \bigcap \varepsilon_2)\right)(k)\right\}$$

$$\Rightarrow (\varepsilon_1 \bigcap \varepsilon_2)(m) \le \max\{0.5, 0.7\} = 0.7.$$

But $(\varepsilon_1 \bigcap \varepsilon_2)(m) = 0.8 \le 0.7$, which is absurd.

Furthermore,

$$\left(\zeta_1 \bigcup \zeta_2\right)(k \odot m) \ge \min\left\{\left(\zeta_1 \bigcup \zeta_2\right)(k), \zeta(_1 \bigcup \zeta_2)(m)\right\}$$

$$\Rightarrow \left(\zeta_1 \bigcup \zeta_2\right)(j) \ge \min\{0.6, 0.7\} = 0.6$$

But $\left(\zeta_1 \bigcup \zeta_2\right)(j) = 0.5 \ge 0.6$, which is ridiculous as well.

Therefore, R-union $(C_1 \bigcup C_2)_R = \left(\bar{\omega}_1 \bigcup \bar{\omega}_2, \omega_1 \bigcap \omega_2\right)$ need not be a cubic intuitionistic linear space.

(ii) Let $\bar{\omega}_1$, $\bar{\omega}_2$, ω_1 and ω_2 be as in (i)

Define $\left(\bar{\omega}_1 \bigcap \bar{\omega}_2\right)$ by $\left(\bar{\omega}_1 \bigcap \bar{\omega}_2\right)$ $(x) = \left\{(\bar{\rho}_1 \bigcap \bar{\rho}_2)(x), (\bar{\sigma}_1 \bigcup \bar{\sigma}_2(x)\right\}$ all $x \in X$, where,

$$(\bar{\rho}_1 \bigcap \bar{\rho}_2)(x) = \min\{\bar{\rho}_1(x),\ \bar{\rho}_2(x)\} \text{and} (\bar{\sigma}_1 \bigcup \bar{\sigma}_2)(x) = \max\{\bar{\sigma}_1(x)\bar{\sigma}_2(x)\}$$

So,

$$(\bar{\rho}_1 \bigcap \bar{\rho}_2)(i) = [0.7,\ 0.9],$$

$$(\bar{\rho}_1 \bigcap \bar{\rho}_2)(j) = [0.2,\ 0.3],$$

$$(\bar{\rho}_1 \bigcap \bar{\rho}_2)(k) = [0.3,\ 0.45],$$

$$(\bar{\rho}_1 \bigcap \bar{\rho}_2)(m) = [0.5,\ 0.65] \text{ and}$$

$$(\bar{\sigma}_1 \bigcup \bar{\sigma}_2)(i) = [0.3,\ 0.4],$$

$$(\bar{\sigma}_1 \bigcup \bar{\sigma}_2)(j) = [0.9,\ 1],$$

$$(\bar{\sigma}_1 \bigcup \bar{\sigma}_2)(k) = [0.9,\ 1],$$

$$(\bar{\sigma}_1 \bigcup \bar{\sigma}_2)(m) = [0.6,\ 0.7].$$

Observe that $\left(\bar{\omega}_1 \bigcap \bar{\omega}_2\right)$ is an interval-valued intuitionistic fuzzy linear space of X.

Also, we prove that example in (i), $\left(\omega_1 \bigcap \omega_2\right)$ is not an intuitionistic fuzzy linear space.

Hence, the P-intersection $(C_1 \bigcap C_2)_P = (\bar{\omega}_1 \bigcap \bar{\omega}_2,\ \omega_1 \bigcap \omega_2)$ need not be a cubic intuitionistic linear space.

(iii) Let $\bar{\omega}_1$, $\bar{\omega}_2$, ω_1 and ω_2 be as in (i)

Moreover, we prove example in (i), $\bar{\omega}_1 \bigcup \bar{\omega}_2$ is not an interval-valued intuitionistic fuzzy linear space.

$\left(\omega_1 \bigcup \omega_2\right)(x) = \left\{(\varepsilon_1 \bigcup \varepsilon_2)(x)\ ,\left(\zeta_1 \bigcap \zeta_2\right)(x)\right\}$ for all $x \in X$, where

$$(\varepsilon_1 \bigcup \varepsilon_2)(x) = \max\{\varepsilon_1(x), \varepsilon_2(x)\} \text{and} \left(\zeta_1 \bigcap \zeta_2\right)(x) \min\{\zeta_1(x)\zeta_2((x)\}$$

Then,

$$\left(\varepsilon_1 \bigcup \varepsilon_2\right)(i) = 0.4,$$

$$\left(\varepsilon_1 \bigcup \varepsilon_2\right)(j) = 0.8 \left(\varepsilon_1 \bigcup \varepsilon_2\right)(k) = 0.91,$$

$$\left(\varepsilon_1 \bigcup \varepsilon_2\right)(m) = 0.91 \text{and}$$

$$\left(\zeta_1 \bigcap \zeta_2\right)(i) = 0.9,$$

$$\zeta_1 \bigcap \zeta_2(j) = 0.4 \text{and} \left(\zeta_1 \bigcap \zeta_2\right)(k) = 0.5,$$

$$\left(\zeta_1 \bigcap \zeta_2\right)(m) = 0.6$$

By verification, it can be seen that $\omega_1 \bigcup \omega_2$ is an intuitionistic fuzzy linear space.

Thus, P-union, $\left(C_1 \bigcup C_2\right)_P = \left(\bar{\omega}_1 \bigcup \bar{\omega}_2, \omega_1 \bigcup \omega_2\right)$ is not a cubic intuitionistic linear space.

REFERENCES

Bag, T., and S.K. Samanta. 2005. Fuzzy bounded linear operators. *Fuzzy Sets and Systems* 151: 513–547.

Chang, S.C., and J.N. Mordesen. 1994. Fuzzy linear operators and fuzzy normed linear spaces. *Bulletin of the Calcutta Mathematical Society* 86: 429–436.

Cho, Y.J., S. Lin, S.S. Kim, and A. Misiak. 2001. *Theory of 2-Inner Product Spaces*. Nova Science Publishers.

Dimmine, C., S. Gähler, and A. White. 1973. 2-inner product spaces. *Demonstratio Mathematica* 6: 525–536.
Felbin, C. 1992. Finite dimensional fuzzy normed linear spaces. *Fuzzy Sets and Systems* 48: 239–248.
Gähler, S. 1965. Lineare 2-Normierte Räume. *Mathematische Nachrichten* 28: 1–43.
Jun, Y.B., T. Jung., and S. Kim. 2010. Cubic subalgebras and ideals. *The Far East Journal of Mathematical Sciences* 44: 239–250.
Jun, Y.B., T. Jung, and S. Kim. 2011. Cubic subgroups. *Annals of Fuzzy Mathematics and Informatics* 2, no. 1: 9–15.
Jun, Y.B., S. Kim, and O. Yang. 2011. Cubic sets. *Annals of Fuzzy Mathematics and Informatics* 4, no. 1: 83–98.
Katsaras, A.K. 1984. Fuzzy topological vector spaces II. *Fuzzy Sets and Systems* 12: 143–154.
Kohli, J.K., and R. Kumar. 1993. On fuzzy inner product spaces and fuzzy co-inner product spaces. *Fuzzy Sets and Systems* 53, no. 2: 227–232.
Krishna, S.V., and K.K.M. Sharma. 1994. Separation of fuzzy normed linear spaces. *Fuzzy Sets and Systems* 63: 207–217.
Ming, M.A. 1985. Comparison between two definitions of fuzzy normed linear spaces. *Harbin Institute of Technology* Sppl. Math, no. A2: 47–49.
Misiak, A. 1989. n-inner product spaces. *Mathematische Nachrichten* 140: 299–319.
Narayanan, A.L., and S. Vijayabalaji. 2005. Fuzzy n-normed linear space. *International Journal of Mathematics and Mathematical Sciences* 2005, no. 24: 3963–3977.
Thillaigovindan, N., S. Vijayabalaji, and S. Anita Shanthi. 2010. *Fuzzy n- Sppl. Math, No. A2 Linear Space*. Lambert Academic Publishing, Mauritius.
Vijayabalaji, S. 2011. Equivalent fuzzy strong n-inner product space. *International Journal of Open Problems in Computer Science and Mathematics* 4, no. 4: 26–32.
Vijayabalaji, S. 2017. *Cubic n- Normed Linear Spaces*. Lambert Academic Publishing, Maude Avenue, Sunnyvale, USA.
Vijayabalaji, S., S. Anita Shanthi., and N. Thillaigovindan. 2009. Interval-valued fuzzy n-normed linear space. *Journal of Fundamental Sciences* 28, no. 2: 283–293.
Vijayabalaji, S., and S. Sivaramakrishnan. 2015. A cubic set theoretical approach to linearspace. *Abstract and Applied Analysis* 2015: 1–8.
Vijayabalaji, S., and N. Thillaigovindan. 2007a. Fuzzy n- inner product space. *Bulletin of Korean Mathematical Society* 43, no. 3: 447–459.
Vijayabalaji, S., and N. Thillaigovindan. 2007b. Complete fuzzy n-normed linear space. *Journal of Fundamental Sciences* 3, no. 1: 119–126.
Zadeh, L.A. 1965. Fuzzy sets. *Information and Control* 8: 338–353.
Zadeh, L.A. 1975. The concept of a linguistic variable and its application to approximate reasoning I. *Information Sciences* 8: 199–249.

9 Convergence Methods for Double Sequences and Applications in Neutrosophic Normed Spaces

Omer Kişi
Bartın University

CONTENTS

9.1 INTRODUCTION

Statistical convergence was firstly examined by Henry Fast (1951). In the wake of the study of ideal convergence (Kostyrko et al. 2000), there has been comprehensive research to discover applications and summability studies of the classical theories. Ideal convergence became a notable topic in summability theory after the research of Dems 2004, Kostyrko et al. 2005, Nabiev et al. 2007, Das and Ghosal 2010, Das et al. 2011, Savaş and Das 2011, Gürdal 2012, Yamancı and Gürdal 2014a, b and Nabiev et al. 2020.

Fuzzy Sets (FSs) were firstly given by Lotfi A. Zadeh (1965). The publication of the paper affected deeply all the scientific fields. This notion is significant for real-life conditions but has no adequate solution to some problems. Such problems lead to original quests.

Intuitionistic fuzzy sets (IFSs) originated from Krassimir Atanassov (1986). Atanassov et al. (2002) utilized this concept in decision-making problems. Ivan Kramosil and Jiří Michálek investigated fuzzy metric space (FMS) utilizing the concepts fuzzy and probabilistic metric space (Kramosil and Michalek 1975). Moreover, FMS has been used in practical research studies, for example, decision-making, fixed point theory, and medical imaging. Jong Hyeok Park (2004) generalized FMSs and defined IF metric space (IFMS). Park utilized the study of George and

Veeramani (1994) and thought of using t-norm and t-conorm to the FMS meantime describing IFMS and investigating its fundamental properties.

Reza Saadati and Jin Han Park initially examined the properties of intuitionistic fuzzy normed space (IFNS) (Saadati and Park 2006). Notable results on this topic can be found in the research of (Karakuş et al. 2008, Mursaleen and Mohiuddine 2009a, b, c, Mohiuddine 2009, Mursaleen et al, 2010, Savaş and Gürdal 2014, 2015, Debnath 2012, 2015, 2016, Debnath and Sen 2014a, b, Konwar and Debnath 2017, Konwar et al. 2018).

Studies on neutrosophy, neutrosophic set, neutrosophic logic, etc. created neutrosophic sets and systems.

The concept neutrosophy implies impartial knowledge of thought and then neutral describes the basic difference between neutral, fuzzy, intuitive fuzzy set, and logic. Neutrosophic logic was firstly initiated by Florent in Smarandache (1995). It is a logic where each proposition is determined to have a degree of truth (T), falsity (F), and indeterminacy (I). A Neutrosophic set (NS) is determined as a set where every component of the universe has a degree of T, F, and I.

Uncertainty is based on the belongingness degree in IFSs, whereas the uncertainty in NS is considered independently from T and F values. Since no limitations among the degree of T, F, I, NSs are actually more general than IFS.

Neutrosophic soft linear spaces (NSLSs) were investigated by Tuhin Bera and Nirmal K. Mahapatra (Bera and Mahapatra 2017). Subsequently, in (Bera and Mahapatra 2018), the concept neutrosophic soft normed linear (NSNLS) was defined, and the features of (NSNLS) were examined.

Kirişci and Şimşek (2020a) defined a new concept known as neutrosophic metric space (NMS) with continuous t-norms and continuous t-conorms. Some notable features of NMS have been examined.

Neutrosophic normed space (NNS) and statistical convergence in NNS have originated from Murat Kirişçi and Necip Şimşek (2020b). Some notable results on this topic can be found in the research of Wang et al. 2010, Yager 2013, Ye 2014, Şimşek and Kirişci 2019, Kirişçi et al. 2020.

The purpose of this chapter is to present and study some recent development in NNS. This chapter provides an overview on NNS. First, we define statistical Cauchy and statistical convergence for double sequences in NNS and prove that every NNS is statistically complete. Additionally, we examine ideal convergence of sequences in NNS. The chapter also aims to analyze studies which criticize I_2^*-convergence, I_2-Cauchy and I_2^*-Cauchy for double sequences in NNS. The fundamental properties of these concepts with regards to NNS are investigated. The results of the chapter are expected to be a source for researchers in the areas of convergence methods for sequences and applications in NNS.

9.2 MAIN RESULTS

Definition 9.1

Take a NNS Ω. (z_{jk}) is called to be statistically convergent sequence to $\rho \in F$ with regard to NN $\mathcal{N}$, if for each $\mathfrak{Z} > 0$ and $\varepsilon > 0$,

$$\delta_2\left(\left\{(j,k)\in\mathbb{N}\times\mathbb{N}:\mathcal{G}\left(z_{jk}-\rho,\mathfrak{z}\right)\leq 1-\varepsilon \text{ or } \mathcal{B}\left(z_{jk}-\rho,\mathfrak{z}\right)\geq\varepsilon,\Upsilon\left(z_{jk}-\rho,\mathfrak{z}\right)\geq\varepsilon\right\}\right)=0,$$

or equivalently

$$\lim_{p,r\to\infty}\frac{1}{pr}\left|\left\{j\leq p,k\leq r:\mathcal{G}\left(z_{jk}-\rho,\mathfrak{z}\right)\leq 1-\varepsilon \text{ or } \mathcal{B}\left(z_{jk}-\rho,\mathfrak{z}\right)\geq\varepsilon,\Upsilon\left(z_{jk}-\rho,\mathfrak{z}\right)\geq\varepsilon\right\}\right|=0.$$

Hence, we indicate $S_{2(\mathcal{N})}-\lim z_{jk}=\rho$ or $z_{jk}\to\rho\left(S_{2(\mathcal{N})}\right)$. The set of S_2 C-NN shall be demonstrated by $S_{2(\mathcal{N})}$.

Definition 9.2

A sequence $\left(z_{jk}\right)$ is called to be statistically Cauchy (S_2 Ca-NN) with regard to NN $\mathcal{N}$ if for each $\mathfrak{z}>0$ and $\varepsilon>0$, there exist $T=T(\varepsilon)$ and $R=R(\varepsilon)$ such that for all, $w\geq T,k,q\geq R,$

$$\delta_2\left\{(j,k)\in\mathbb{N}\times\mathbb{N}:\mathcal{G}\left(z_{jk}-z_{wq},\mathfrak{z}\right)\leq 1-\varepsilon \text{ or } \mathcal{B}\left(z_{jk}-z_{wq},\mathfrak{z}\right)\geq\varepsilon,\Upsilon\left(z_{jk}-z_{wq},\mathfrak{z}\right)\geq\varepsilon\right\}=0.$$

Definition 9.3

If every S_2 Ca-NN is S_2 C-NN, then NNS Ω is statistically complete.

Theorem 9.1

Every NNS Ω is statistically complete.

Proof: Let $\left(z_{jk}\right)$ be S_2 Ca-NN but not S_2 C-NN. Then there exist w and q such that the set $\delta_2\left(T\left((\varepsilon,\mathfrak{z})\right)\right)=0$, where

$$T\left((\varepsilon,\mathfrak{z})\right)$$
$$=\left\{(j,k)\in\mathbb{N}\times\mathbb{N}:\mathcal{G}\left(z_{jk}-z_{wq},\mathfrak{z}\right)\leq 1-\varepsilon \text{ or } \mathcal{B}\left(z_{jk}-z_{wq},\mathfrak{z}\right)\geq\varepsilon,\Upsilon\left(z_{jk}-z_{wq},\mathfrak{z}\right)\geq\varepsilon\right\}$$

and also $\delta_2\left(P\left((\varepsilon,\mathfrak{z})\right)\right)=0$, where

$$P\left((\varepsilon,\mathfrak{z})\right)$$
$$=\left\{(j,k)\in\mathbb{N}\times\mathbb{N}:\mathcal{G}\left(z_{jk}-\rho,\frac{\mathfrak{z}}{2}\right)>1-\varepsilon \text{ and } \mathcal{B}\left(z_{jk}-\rho,\frac{\mathfrak{z}}{2}\right)<\varepsilon,\Upsilon\left(z_{jk}-\rho,\frac{\mathfrak{z}}{2}\right)<\varepsilon\right\}.$$

Since

$$\mathcal{G}\left(z_{jk}-z_{wq},\mathfrak{z}\right)\geq 2\mathcal{G}\left(z_{jk}-\rho,\frac{\mathfrak{z}}{2}\right)>1-\varepsilon,$$

and

$$\mathcal{B}(z_{jk} - z_{wq}, \mathfrak{z}) \leq 2\,\mathcal{B}\left(z_{jk} - \rho, \frac{\mathfrak{z}}{2}\right) < \varepsilon, \quad \Upsilon(z_{jk} - z_{wq}, \mathfrak{z}) \leq 2\,\Upsilon\left(z_{jk} - \rho, \frac{\mathfrak{z}}{2}\right) < \varepsilon,$$

if $\mathcal{G}\left(z_{jk} - \rho, \frac{\mathfrak{z}}{2}\right) > \frac{(1-\varepsilon)}{2}$ and $\mathcal{B}\left(z_{jk} - \rho, \frac{\mathfrak{z}}{2}\right) < \frac{\varepsilon}{2}$, $\Upsilon\left(z_{jk} - \rho, \frac{\mathfrak{z}}{2}\right) < \frac{\varepsilon}{2}$. Therefore

$$\delta_2\left(\left\{(j,k) \in \mathbb{N} \times \mathbb{N} : \mathcal{G}(z_{jk} - z_{wq}, \mathfrak{z}) > 1 - \varepsilon \text{ and } \mathcal{B}(z_{jk} - z_{wq}, \mathfrak{z}) < \varepsilon, \Upsilon(z_{jk} - z_{wq}, \mathfrak{z}) < \varepsilon\right\}\right)$$
$$= 0,$$

that is, $\delta_2\left(T^c\left((\varepsilon, \mathfrak{z})\right)\right) = 0$ and hence $\delta_2\left(T\left((\varepsilon, \mathfrak{z})\right)\right) = 1$, which leads to contradiction. As a result (z_{jk}) have to be S_2 C-NN. We conclude that every NNS Ω is statistically complete.

Definition 9.4

Take a NNS Ω and nontrivial ideal I_2.(z_{jk}) is called to be I_2-convergent sequence to $\rho \in F$ with regard to NN $\mathcal{N}$ (I_2 C-NN), if for each $\mathfrak{z} > 0$ and $\varepsilon > 0$,

$$\left\{(j,k) \in \mathbb{N} \times \mathbb{N} : \mathcal{G}(z_{jk} - \rho, \mathfrak{z}) \leq 1 - \varepsilon \text{ or } \mathcal{B}(z_{jk} - \rho, \mathfrak{z}) \geq \varepsilon, \Upsilon(z_{jk} - \rho, \mathfrak{z}) \geq \varepsilon\right\} \in I_2.$$

It is indicated by $I_{2(\mathcal{N})} - \lim z_{jk} = \rho$ or $z_{jk} \to \rho\left(I_{2(N)}\right)$. The set of I_2 C-NN will be demonstrated by $I_{2(\mathcal{N})}$.

Lemma 9.1

Take a NNS Ω. Then, for each $\mathfrak{z} > 0$ and $\varepsilon > 0$, the following situations are equivalent:

1. $I_{2(\mathcal{N})} - \lim z_{jk} = \rho$,
2. $\left\{(j,k) \in \mathbb{N} \times \mathbb{N} : \mathcal{G}(z_{jk} - \rho, \mathfrak{z}) \leq 1 - \varepsilon\right\} \in I_2$ and
   $\left\{(j,k) \in \mathbb{N} \times \mathbb{N} : \mathcal{B}(z_{jk} - \rho, \mathfrak{z}) \geq \varepsilon\right\} \in I_2$, $\left\{(j,k) \in \mathbb{N} \times \mathbb{N} : \Upsilon(z_{jk} - \rho, \mathfrak{z}) \geq \varepsilon\right\} \in I_2$.
3. $\left\{(j,k) \in \mathbb{N} \times \mathbb{N} : \mathcal{G}(z_{jk} - \rho, \mathfrak{z}) > 1 - \varepsilon \text{ and } \mathcal{B}(z_{jk} - \rho, \mathfrak{z}) < \varepsilon, \Upsilon(z_{jk} - \rho, \mathfrak{z}) < \varepsilon\right\}$
   $\in F(I_2)$.
4. $\left\{(j,k) \in \mathbb{N} \times \mathbb{N} : \mathcal{G}(z_{jk} - \rho, \mathfrak{z}) > 1 - \varepsilon\right\} \in F(I_2)$and
   $\left\{(j,k) \in \mathbb{N} \times \mathbb{N} : \mathcal{B}(z_{jk} - \rho, \mathfrak{z}) < \varepsilon\right\} \in F(I_2)$,
   $\left\{(j,k) \in \mathbb{N} \times \mathbb{N} : \Upsilon(z_{jk} - \rho, \mathfrak{z}) < \varepsilon\right\} \in F(I_2)$.

5. $I_{2(\mathcal{N})}-\lim\mathcal{G}(z_{jk}-\rho,\mathfrak{z})=1$ and $I_{2(\mathcal{N})}-\lim\mathcal{B}(z_{jk}-\rho,\mathfrak{z})$

$=0, I_{2(\mathcal{N})}-\lim\Upsilon(z_{jk}-\rho,\mathfrak{z})=0.$

Theorem 9.2

If (z_{jk}) is I_2 C-NN, then $I_{2(\mathcal{N})}-$ limit is unique.

Proof: Presume that $I_{2(\mathcal{N})}-\lim z_{jk}=\rho_1$ and $I_{2(\mathcal{N})}-\lim z_{jk}=\rho_2$. Select $\varepsilon>0$. Then, for a given $\mathfrak{r}>0,(1-\mathfrak{r})\circ(1-\mathfrak{r})>1-\varepsilon$ and $\mathfrak{r}*\mathfrak{r}<\varepsilon$. For any $\mathfrak{z}>0$, let's denote the following sets:

$$K_{\mathcal{G}1}(\mathfrak{r},\mathfrak{z}):=\left\{(j,k)\in\mathbb{N}\times\mathbb{N}:\mathcal{G}\left(z_{jk}-\rho_1,\frac{\mathfrak{z}}{2}\right)\leq 1-\mathfrak{r}\right\},$$

$$K_{\mathcal{B}1}(\mathfrak{r},\mathfrak{z}):=\left\{(j,k)\in\mathbb{N}\times\mathbb{N}:\mathcal{B}\left(z_{jk}-\rho_1,\frac{\mathfrak{z}}{2}\right)\geq \mathfrak{r}\right\},$$

$$K_{\Upsilon_1}(\mathfrak{r},\mathfrak{z}):=\left\{(j,k)\in\mathbb{N}\times\mathbb{N}:\Upsilon\left(z_{jk}-\rho_1,\frac{\mathfrak{z}}{2}\right)\geq \mathfrak{r}\right\}$$

and

$$K_{\mathcal{G}2}(\mathfrak{r},\mathfrak{z}):=\left\{(j,k)\in\mathbb{N}\times\mathbb{N}:\mathcal{G}\left(z_{jk}-\rho_2,\frac{\mathfrak{z}}{2}\right)\leq 1-\mathfrak{r}\right\},$$

$$K_{\mathcal{B}2}(\mathfrak{r},\mathfrak{z}):=\left\{(j,k)\in\mathbb{N}\times\mathbb{N}:\mathcal{B}\left(z_{jk}-\rho_2,\frac{\mathfrak{z}}{2}\right)\geq \mathfrak{r}\right\},$$

$$K_{\Upsilon_2}(\mathfrak{r},\mathfrak{z}):=\left\{(j,k)\in\mathbb{N}\times\mathbb{N}:\Upsilon\left(z_{jk}-\rho_2,\frac{\mathfrak{z}}{2}\right)\geq \mathfrak{r}\right\}.$$

Since $I_{2(\mathcal{N})}-\lim z_{jk}=\rho_1$, we have $K_{\mathcal{G}1}(\mathfrak{r},\mathfrak{z}), K_{\mathcal{B}1}(\mathfrak{r},\mathfrak{z})$ and $K_{\Upsilon_1}(\mathfrak{r},\mathfrak{z})\in I_2$. Also, using $I_{2(\mathcal{N})}-\lim z_{jk}=\rho_2$, we have $K_{\mathcal{G}2}(\mathfrak{r},\mathfrak{z}), K_{\mathcal{B}_2}(\mathfrak{r},\mathfrak{z})$, and $K_{\Upsilon_2}(\mathfrak{r},\mathfrak{z})\in I_2$. Now, consider

$$K_{\mathcal{N}}((\mathfrak{r},\mathfrak{z})):=\{K_{\mathcal{G}1}(\mathfrak{r},\mathfrak{z})\cup K_{\mathcal{G}2}(\mathfrak{r}\ \mathfrak{z})\}\cap\{K_{\mathcal{B}1}(\mathfrak{r}\ \mathfrak{z})\cup K_{\mathcal{B}2}(\mathfrak{r},\mathfrak{z})\}\cap$$
$$\{K_{\Upsilon_1}(\mathfrak{r},\mathfrak{z})\cup K_{\Upsilon_2}(\mathfrak{r},\mathfrak{z})\}.$$

Then, $K_{\mathcal{N}}((\mathfrak{r},\mathfrak{z}))\in I_2$, which means that $\varnothing\neq K^c_{\mathcal{N}}((\mathfrak{r},\mathfrak{z}))\in F(I_2)$. If $(j,k)\in K^c_{\mathcal{N}}((\mathfrak{r},\mathfrak{z}))$, then we have three possible situations. That is, $(j,k)\in K^c_{\mathcal{G}1}((\mathfrak{r},\mathfrak{z}))\cap K^c_{\mathcal{G}2}((\mathfrak{r},\mathfrak{z}))$, $(j,k)\in K^c_{\mathcal{B}1}((\mathfrak{r},\mathfrak{z}))\cap K^c_{\mathcal{B}2}((\mathfrak{r},\mathfrak{z}))$ or $(j,k)\in K^c_{\Upsilon_1}((\mathfrak{r},\mathfrak{z}))\cap K^c_{\Upsilon_2}((\mathfrak{r},\mathfrak{z}))$. First, consider that $(j,k)\in K^c_{\mathcal{G}1}((\mathfrak{r},\mathfrak{z}))\cap K^c_{\mathcal{G}2}((\mathfrak{r},\mathfrak{z}))$. Then, we obtain

$$\mathcal{G}(\rho_1-\rho_2,\mathfrak{z})\geq\mathcal{G}\left(z_{jk}-\rho_1,\frac{\mathfrak{z}}{2}\right)\circ\mathcal{G}\left(z_{jk}-\rho_2,\frac{\mathfrak{z}}{2}\right)>(1-\mathfrak{r})\circ(1-\mathfrak{r})1-\varepsilon.$$

For arbitrary $\varepsilon > 0$, we get $\mathcal{G}(\rho_1 - \rho_2, \mathfrak{z}) = 1$ for all $\mathfrak{z} > 0$, which yields $\rho_1 = \rho_2$. https://www.powerthesaurus.org/at_the_same_time/synonyms At the same time, if we take $(j,k) \in K^c_{\mathcal{B}_1}((\mathfrak{r},\mathfrak{z})) \cap K^c_{\mathcal{B}_2}((\mathfrak{r},\mathfrak{z}))$, then we can evidence

$$\mathcal{B}(\rho_1 - \rho_2, \mathfrak{z}) \le \mathcal{B}\left(z_{jk} - \rho_1, \frac{\mathfrak{z}}{2}\right) * \mathcal{B}\left(z_{jk} - \rho_2, \frac{\mathfrak{z}}{2}\right) < \mathfrak{r} * \mathfrak{r} < \varepsilon.$$

Therefore, we can see that $\mathcal{B}(\rho_1 - \rho_2, \mathfrak{z}) < \varepsilon$. For all $\mathfrak{z} > 0$, we acquire $\mathcal{B}(\rho_1 - \rho_2, \mathfrak{z}) = 0$, which gives that $\rho_1 = \rho_2$. Again, for the situation $(j,k) \in K^c_{\Upsilon_1}((\mathfrak{r},\mathfrak{z})) \cap K^c_{\Upsilon_2}((\mathfrak{r},\mathfrak{z}))$, then we have

$$\Upsilon(\rho_1 - \rho_2, \mathfrak{z}) \le \Upsilon\left(z_{jk} - \rho_1, \frac{\mathfrak{z}}{2}\right) * \Upsilon\left(z_{jk} - \rho_2, \frac{\mathfrak{z}}{2}\right) < \mathfrak{r} * \mathfrak{r} < \varepsilon.$$

For all $\mathfrak{z} > 0$, we get $\Upsilon(\rho_1 - \rho_2, \mathfrak{z}) = 0$. Thus, $\rho_1 = \rho_2$. In all situations, we deduce that the $I_{2(\mathcal{N})}$ – limit is unique.

Theorem 9.3

If $\mathcal{N} - \lim z_{jk} = \rho$ for NNS Ω, then $I_{2(\mathcal{N})} - \lim z_{jk} = \rho$.

Proof: Presume that $\mathcal{N} - \lim z_{jk} = \rho$. Then, for each $\mathfrak{z} > 0$, and $\varepsilon > 0$, there exists $T > 0$ such that $\mathcal{G}(z_{jk} - \rho, \mathfrak{z}) > 1 - \varepsilon$ and $\mathcal{B}(z_{jk} - \rho, \mathfrak{z}) < \varepsilon, \Upsilon(z_{jk} - \rho, \mathfrak{z}) < \varepsilon$ for all $j,k \ge T$. Since the set

$$P(\varepsilon,\mathfrak{z}) := \left\{(j,k) \in \mathbb{N} \times \mathbb{N} : \mathcal{G}(z_{jk} - \rho, \mathfrak{z}) \le 1 - \varepsilon \text{ or } \mathcal{B}(z_{jk} - \rho, \mathfrak{z}) \ge \varepsilon, \Upsilon(z_{jk} - \rho, \mathfrak{z}) \ge \varepsilon\right\}$$

is included in $\{1,2,3,\ldots,T-1\}$ and I_2 is admissible, $P(\varepsilon,\mathfrak{z}) \in I_2$, Hence, $I_{2(\mathcal{N})} - \lim z_{jk} = \rho$.

To show converse, we consider the following:

Example 9.1

Let $(F,.)$ be a NS. For all $p,q \in [0,1]$, take the t-norm (TN) $p \circ q = pq$ and the t-conorm (TC) $p * q = \min\{p+q, 1\}$. Contemplate $\mathcal{G}(z,\mathfrak{z}) = \frac{\mathfrak{z}}{\mathfrak{z}+z}$, $\mathcal{B}(z,\mathfrak{z}) = \frac{z}{\mathfrak{z}+z}$ and $\Upsilon(z,\mathfrak{z}) = \frac{z}{\mathfrak{z}}$ for all $z \in F$ and each $\mathfrak{z} > 0$. Then Ω is a NNS. We select (z_{jk}) by

$$z_{jk} = \begin{cases} jk, & \text{if } j = n^2 \text{ and } k = m^2 \ (n,m \in \mathbb{N}) \\ 0, & \text{otherwise.} \end{cases}$$

Consider

$$P(\varepsilon,\mathfrak{z}) := \left\{(j,k) \in \mathbb{N} \times \mathbb{N} : \mathcal{G}(z_{jk}, \mathfrak{z}) \le 1 - \varepsilon \text{ or } \mathcal{B}(z_{jk}, \mathfrak{z}) \ge \varepsilon, \Upsilon(z_{jk}, \mathfrak{z}) \ge \varepsilon\right\}$$

for any $\mathfrak{z} > 0$ and for each $\varepsilon > 0$. Then, we get

$$P(\varepsilon,\mathfrak{z}) = \left\{(j,k) \in \mathbb{N}\times\mathbb{N} : \frac{\mathfrak{z}}{\mathfrak{z}+z_{jk}} \le 1-\varepsilon \text{ or } \frac{z_{jk}}{\mathfrak{z}+z_{jk}} \ge \varepsilon, \frac{z_{jk}}{\mathfrak{z}} \ge \varepsilon\right\}$$

$$= \left\{(j,k) \in \mathbb{N}\times\mathbb{N} : \quad z_{jk} \ge \frac{\mathfrak{z}\varepsilon}{1-\varepsilon} > 0 \text{ or } z_{jk} \ge \mathfrak{z}\varepsilon > 0\right\}$$

$$= \left\{(j,k) \in \mathbb{N}\times\mathbb{N} : \; z_{jk} = jk\right\} = \left\{(j,k) \in \mathbb{N}\times\mathbb{N} : j = n^2 \text{ and } k = m^2, \; (n,m \in \mathbb{N})\right\}$$

which is finite set. So, $P(\varepsilon,\mathfrak{z}) \in I_2$, where I_2 is admissible ideal. As a result, $z_{jk} \to 0\left(I_{2(\mathcal{N})}\right)$.

On the other side, $z_{jk} \nrightarrow 0(\mathcal{N})$, since

$$\mathcal{G}\left(z_{jk},\mathfrak{z}\right) = \frac{\mathfrak{z}}{\mathfrak{z}+\left|z_{jk}\right|} = \begin{cases} \dfrac{\mathfrak{z}}{\mathfrak{z}+jk}, & \text{if } j = n^2 \text{ and } k = m^2 (n,m \in \mathbb{N}) \\ 1, & \text{otherwise;} \end{cases}$$

and

$$\mathcal{B}\left(z_{jk},\mathfrak{z}\right) = \frac{\left|z_{jk}\right|}{\mathfrak{z}+\left|z_{jk}\right|} = \begin{cases} \dfrac{jk}{\mathfrak{z}+jk}, & \text{if } j = n^2 \text{ and } k = m^2 (n,m \in \mathbb{N}) \\ 0, & \text{otherwise;} \end{cases}$$

$$\Upsilon\left(z_{jk},\mathfrak{z}\right) = \frac{\left|z_{jk}\right|}{\mathfrak{z}} = \begin{cases} \dfrac{jk}{\mathfrak{z}}, & \text{if } j = n^2 \text{ and } k = m^2 (n,m \in \mathbb{N}) \\ 0, & \text{otherwise.} \end{cases}$$

we obtain $\mathcal{G}\left(z_{jk},\mathfrak{z}\right) \le 1$, and $\mathcal{B}\left(z_{jk},\mathfrak{z}\right) \ge 0, \Upsilon\left(z_{jk},\mathfrak{z}\right) \ge 0$. This concludes the proof.

Theorem 9.4

Let Ω be a NNS. Then,

1. If $I_{2(\mathcal{N})} - \lim y_{jk} = \rho_1$ and $I_{2(\mathcal{N})} - \lim z_{jk} = \rho_2$, then $I_{2(\mathcal{N})} - \lim\left(y_{jk} + z_{jk}\right) = \rho_1 + \rho_2$.
2. If $I_{2(\mathcal{N})} - \lim z_{jk} = \rho$, then $I_{2(\mathcal{N})} - \lim \beta z_{jk} = \beta\rho$.

Proof: (1) Let $I_{2(\mathcal{N})} - \lim y_{jk} = \rho_1$ and $I_{2(\mathcal{N})} - \lim z_{jk} = \rho_2$. Choose $\mathfrak{r} > 0$. Then, for a given $\varepsilon > 0, (1-\mathfrak{r}) \circ (1-\mathfrak{r}) > 1-\varepsilon$ and $\mathfrak{r} * \mathfrak{r} < \varepsilon$. Then, for any $\mathfrak{z} > 0$, we select the following sets: $K_{\mathcal{G}_1}(\mathfrak{r},\mathfrak{z}), K_{\mathcal{B}_1}(\mathfrak{r},\mathfrak{z}), K_{\Upsilon_1}(\mathfrak{r},\mathfrak{z}), K_{\mathcal{G}_2}(\mathfrak{r},\mathfrak{z}), K_{\mathcal{B}_2}(\mathfrak{r},\mathfrak{z}), K_{\Upsilon_2}(\mathfrak{r},\mathfrak{z})$ as above. Since $I_{2(\mathcal{N})} - \lim y_{jk} = \rho_1$, we have $K_{\mathcal{G}_1}(\mathfrak{r},\mathfrak{z}), K_{\mathcal{B}_1}(\mathfrak{r},\mathfrak{z}), K_{\Upsilon_1}(\mathfrak{r},\mathfrak{z}) \in I_2$. Further, using $I_{2(\mathcal{N})} - \lim z_{jk} = \rho_2 K_{\mathcal{G}_2}(\mathfrak{r},\mathfrak{z}), K_{\mathcal{B}_2}(\mathfrak{r},\mathfrak{z}), K_{\Upsilon_2}(\mathfrak{r},\mathfrak{z}) \in I_2$. Now, consider

$$K_{\mathcal{N}}((\mho,\mathfrak{z})):=\left\{K_{\mathcal{G}_1}(\mho,\mathfrak{z})\bigcup K_{\mathcal{G}_2}(\mho,\mathfrak{z})\right\}\bigcap\left\{K_{\mathcal{B}_1}(\mho,\mathfrak{z})\bigcup K_{\mathcal{B}_2}(\mho,\mathfrak{z})\right\}\bigcap$$
$$\left\{K_{\Upsilon_1}(\mho,\mathfrak{z})\bigcup K_{\Upsilon_2}(\mho,\mathfrak{z})\right\}.$$

Then, $K_{\mathcal{N}}((\mho,\mathfrak{z}))\in I_2$, which yields that $\varnothing\neq K_{\mathcal{N}}^c((\mho,\mathfrak{z}))\in F(I_2)$. Now we have to indicate that

$$K_{\mathcal{N}}^c((\mho,\mathfrak{z}))\subset\Big\{(j,k)\in\mathbb{N}\times\mathbb{N}:\mathcal{G}\big((y_{jk}+z_{jk})-(\rho_1+\rho_2),\mathfrak{z}\big)>1-\varepsilon \text{ and}$$
$$\mathcal{B}\big((y_{jk}+z_{jk})-(\rho_1+\rho_2),\mathfrak{z}\big)<\varepsilon,\Upsilon\big((y_{jk}+z_{jk})-(\rho_1+\rho_2),\mathfrak{z}\big)<\varepsilon\Big\}.$$

If $(j,k)\in K_{\mathcal{N}}^c((\mho,\mathfrak{z}))$, then we get $\left(y_{jk}-\rho_1,\frac{\mathfrak{z}}{2}\right)>1-\mho,\mathcal{G}\left(z_{jk}-\rho_2,\frac{\mathfrak{z}}{2}\right)>1-\mho$, $\mathcal{B}\left(y_{jk}-\rho_1,\frac{\mathfrak{z}}{2}\right)<\mho,B\left(z_{jk}-\rho_2,\frac{\mathfrak{z}}{2}\right)<\mho,Y\left(y_{jk}-\rho_1,\frac{\mathfrak{z}}{2}\right)<\mho$ and $\mathcal{Y}\left(z_{jk}-\rho_2,\frac{\mathfrak{z}}{2}\right)<\mho$. Therefore,

$$\mathcal{G}\big((y_{jk}+z_{jk})-(\rho_1+\rho_2),\mathfrak{z}\big)\geq\mathcal{G}\left(y_{jk}-\rho_1,\frac{\mathfrak{z}}{2}\right)\circ\mathcal{G}\left(z_{jk}-\rho_2,\frac{\mathfrak{z}}{2}\right)>(1-\mho)\circ(1-\mho)>1-\varepsilon,$$

$$\mathcal{B}\big((y_{jk}+z_{jk})-(\rho_1+\rho_2),\mathfrak{z}\big)\leq\mathcal{B}\left(y_{jk}-\rho_1,\frac{\mathfrak{z}}{2}\right)*\mathcal{B}\left(z_{jk}-\rho_2,\frac{\mathfrak{z}}{2}\right)<\mho*\mho<\varepsilon$$

and

$$\Upsilon\big((y_{jk}+z_{jk})-(\rho_1+\rho_2),\mathfrak{z}\big)\leq\Upsilon\left(y_{jk}-\rho_1,\frac{\mathfrak{z}}{2}\right)*\Upsilon\left(z_{jk}-\rho_2,\frac{\mathfrak{z}}{2}\right)<\mho*\mho<\varepsilon.$$

This gives

$$K_{\mathcal{N}}^c((\mho,\mathfrak{z}))\subset\Big\{(j,k)\in\mathbb{N}\times\mathbb{N}:\mathcal{G}\big((y_{jk}+z_{jk})-(\rho_1+\rho_2),\mathfrak{z}\big)>1-\varepsilon \text{ and}$$
$$\mathcal{B}\big((y_{jk}+z_{jk})-(\rho_1+\rho_2),\mathfrak{z}\big)<\varepsilon,\Upsilon\big((y_{jk}+z_{jk})-(\rho_1+\rho_2),\mathfrak{z}\big)<\varepsilon\Big\}.$$

Since $K_{\mathcal{N}}^c((\mho,\mathfrak{z}))\in F(I_2), I_{2(\mathcal{N})}-\lim(y_{jk}+z_{jk})=\rho_1+\rho_2$.

This is clear for $\beta=0$. Let $\beta\neq 0$. Then, for a given $\mathfrak{z}>0$ and $\varepsilon>0$,

$$T(\varepsilon,\mathfrak{z})=\Big\{(j,k)\in\mathbb{N}\times\mathbb{N}:\mathcal{G}(z_{jk}-\rho,\mathfrak{z})>1-\varepsilon \text{ and}$$
$$\mathcal{B}(z_{jk}-\rho,\mathfrak{z})<\varepsilon,\Upsilon(z_{jk}-\rho,\mathfrak{z})<\varepsilon\Big\}\in F(I_2). \quad (9.1)$$

It is adequate to confirm that for each $\mathfrak{z} > 0$ and $\varepsilon > 0$,

$$T(\varepsilon,\mathrm{n}) \subset \{(j,k) \in \mathbb{N} \times \mathbb{N} : \mathcal{G}(\beta z_{jk} - \beta\rho, \mathfrak{z}) > 1 - \varepsilon \text{ and}$$
$$\mathcal{B}(\beta z_{jk} - \beta\rho, \mathfrak{z}) < \varepsilon, \Upsilon(\beta z_{jk} - \beta\rho, \mathfrak{z}) < \varepsilon\}.$$

Let $(j,k) \in T(\varepsilon,\mathfrak{z})$. So, we have

$$\mathcal{G}(z_{jk} - \rho, \mathfrak{z}) > 1 - \varepsilon \text{ and } \mathcal{B}(z_{jk} - \rho, \mathfrak{z}) < \varepsilon, \Upsilon(z_{jk} - \rho, \mathfrak{z}) < \varepsilon.$$

Hence, we get

$$\mathcal{G}(\beta z_{jk} - \beta\rho, \mathfrak{z}) = \mathcal{G}\left(z_{jk} - \rho, \frac{\mathfrak{z}}{|\beta|}\right) \geq \mathcal{G}(z_{jk} - \rho, \mathfrak{z}) \circ \mathcal{G}\left(0, \frac{\mathfrak{z}}{|\beta|} - \mathfrak{z}\right)$$
$$= \mathcal{G}(z_{jk} - \rho, \mathfrak{z}) \circ 1 = \mathcal{G}(z_{jk} - \rho, \mathfrak{z}) > 1 - \varepsilon,$$

$$\mathcal{B}(\beta z_{jk} - \beta\rho, \mathfrak{z}) = \mathcal{B}\left(z_{jk} - \rho, \frac{\mathfrak{z}}{|\beta|}\right) \leq \mathcal{B}(z_{jk} - \rho, \mathfrak{z}) * \mathcal{B}\left(0, \frac{\mathfrak{z}}{|\beta|} - \mathfrak{z}\right)$$
$$= \mathcal{B}(z_{jk} - \rho, \mathfrak{z}) * 0 = \mathcal{B}(z_{jk} - \rho, \mathfrak{z}) < \varepsilon.$$

In addition,

$$\Upsilon(\beta z_{jk} - \beta\rho, \mathfrak{z}) = \Upsilon\left(z_{jk} - \rho, \frac{\mathfrak{z}}{|\beta|}\right) \leq \Upsilon(z_{jk} - \rho, \mathfrak{z}) * \Upsilon\left(0, \frac{\mathfrak{z}}{|\beta|} - \mathfrak{z}\right)$$
$$= \Upsilon(z_{jk} - \rho, \mathfrak{z}) * 0 = \Upsilon(z_{jk} - \rho, \mathfrak{z}) < \varepsilon.$$

Hence, we get

$$T(\varepsilon,\mathfrak{z}) \subset \{(j,k) \in \mathbb{N} \times \mathbb{N} : \mathcal{G}(\beta z_{jk} - \beta\rho, \mathfrak{z}) > 1 - \varepsilon \text{ and}$$
$$\mathcal{B}(\beta z_{jk} - \beta\rho, \mathfrak{z}) < \varepsilon, \Upsilon(\beta z_{jk} - \beta\rho, \mathfrak{z}) < \varepsilon\},$$

and from equation (9.1) we conclude that $I_{2(\mathcal{N})} - \lim \beta z_{jk} = \beta\rho$.

At the moment, we investigate $I^{*}_{2(\mathcal{N})}$-convergence of double sequences in NNS.

Definition 9.5

A sequence (z_{jk}) is called to be $I^*_{2(\mathcal{N})}$ -convergent to $\rho \in F$ with regard to NN $\mathcal{N}$, if there exists $P = \{(j_m, k_m): j_1 < j_2 < \ldots; k_1 < k_2 < \ldots\} \subset \mathbb{N} \times \mathbb{N}$ such that $P \in F(I_2)$ (i.e. $(\mathbb{N} \times \mathbb{N}) \setminus P \in I_2$) and $\mathcal{N} - \lim_m z_{j_m k_m} = \rho. \otimes$

In addition, it is denoted by $I^*_{2(\mathcal{N})} - \lim z = \rho$.

Theorem 9.5

If $I^*_{2(\mathcal{N})} - \lim z = \rho$, then $I_{2(\mathcal{N})} - \lim z_{jk} = \rho$.

Proof: Presume that $I^*_{2(\mathcal{N})} - \lim z = \rho$,

Then $P = \{(j_m, k_m): j_1 < j_2 < \ldots; k_1 < k_2 < \ldots\} \in F(I_2)$ (i.e. $(\mathbb{N} \times \mathbb{N}) \setminus P = H \in I_2$) such that $\mathcal{N} - \lim_m z_{j_m k_m} = \rho$. But, then for each $\mathfrak{z} > 0$ and $\varepsilon > 0$, there exists $T > 0$ such that $\mathcal{G}(z_{j_m k_m} - \rho, \mathfrak{z}) > 1 - \varepsilon$ and $\mathcal{B}(z_{j_m k_m} - \rho, \mathfrak{z}) < \varepsilon, \Upsilon(z_{j_m k_m} - \rho, \mathfrak{z}) < \varepsilon$ for all $m > T$. Since $\{(j_m, k_m) \in P : \mathcal{G}(z_{j_m k_m} - \rho, \mathfrak{z}) \le 1 - \varepsilon$ or $\mathcal{B}(z_{j_m k_m} - \rho, \mathfrak{z}) \ge \varepsilon, \Upsilon(z_{j_m k_m} - \rho, \mathfrak{z}) \ge \varepsilon\}$ is included in $\{j_1 < j_2 < \ldots < j_{T-1}; k_1 < k_2 < \ldots < k_{T-1}\}$ and the ideal I_2 is admissible, we get $\{(j_m, k_m) \in P : \mathcal{G}(z_{j_m k_m} - \rho, \mathfrak{z}) \le 1 - \varepsilon$ or $\mathcal{B}(z_{j_m k_m} - \rho, \mathfrak{z}) \ge \varepsilon, \Upsilon(z_{j_m k_m} - \rho, \mathfrak{n}) \ge \varepsilon\}$ $\in I_2$. Hence

$$\{(j,k) \in \mathbb{N} \times \mathbb{N} : \mathcal{G}(z_{jk} - \rho, \mathfrak{z}) \le 1 - \varepsilon \text{ or } \mathcal{B}(z_{jk} - \rho, \mathfrak{z}) \ge \varepsilon, \Upsilon(z_{jk} - \rho, \mathfrak{z}) \ge \varepsilon\} \subseteq H \bigcup$$
$$\{j_1 < j_2 < \ldots < j_{T-1}; k_1 < k_2 < \ldots < k_{T-1}\} \in I_2$$

for all $\mathfrak{z} > 0$ and $\varepsilon > 0$. Therefore, we finalize that $I_{2(\mathcal{N})} - \lim z_{jk} = \rho$.

Remark 9.1

The example below denotes that the converse of Theorem 9.5 does not have to be true.

Example 9.2

Take $(\mathbb{R}, .)$. For all $p, q \in [0,1]$, think the TN $p \circ q = pq$ and the TC $p * q = \min\{p + q, 1\}$. For all $z \in F$ and every $\varepsilon > 0$., we consider $\mathcal{G}(z, \mathfrak{z}), \mathcal{B}(z, \mathfrak{z})$ and $\Upsilon(z, \mathfrak{z})$ as Example 9.1. Then, $(\mathbb{R}, \mathcal{N}, \circ, *)$ be NNS. Let $\mathbb{N} \times \mathbb{N} = \bigcup_{i,j} \Delta_{ij}$ be a decomposition of $\mathbb{N} \times \mathbb{N}$ such that, for any $(s,t) \in \mathbb{N} \times \mathbb{N}$, each Δ_{ij} involves infinitely many (i,j)'s, where $i \ge s, j \ge t$ and $\Delta_{ij} \bigcap \Delta_{st} = \varnothing$ for $(i,j) \ne (s,t)$. At the moment, we select a sequence $z_{st} = \frac{1}{ij}$ if $(s,t) \in \Delta_{ij}$. Then

$$\mathcal{G}(z_{st}, \mathfrak{z}) = \frac{\mathfrak{z}}{\mathfrak{z} + z_{st}} \to 1 \text{ and } \mathcal{B}(z_{st}, \mathfrak{z}) = \frac{z_{st}}{\mathfrak{z} + z_{st}} \to 0, \ \Upsilon(z_{st}, \mathfrak{z}) = \frac{z_{st}}{\mathfrak{z}} \to 0 \ \text{ as } s,t \to \infty.$$

Hence, $I_{2(\mathcal{N})} - \lim_{st} z_{st} = 0$.

Assume that $I^*_{2(\mathcal{N})} - \lim_{st} z_{st} = 0$. Then, there exists $P = \{s_1 < s_2 < \ldots; t_1 < t_2 < \ldots\} \subset \mathbb{N} \times \mathbb{N}$ such that $P \in F(I_2)$ and $\mathcal{N} - \lim_j z_{s_j t_j} = 0.$ — Since $P \in F(I_2)$, there is a set $H \in I_2$ such that $P = (\mathbb{N} \times \mathbb{N}) \setminus H$. From the description of I_2, there exists $q \in \mathbb{N}$ such that

$$H \subset \left(\bigcup_{s=1}^{q} \left(\bigcup_{t=1}^{\infty} \Delta_{st} \right) \right) \bigcup \left(\bigcup_{t=1}^{q} \left(\bigcup_{s=1}^{\infty} \Delta_{st} \right) \right).$$

But then $\Delta_{q+1,p+1} \subset P$, and therefore $z_{s_j t_j} = \dfrac{1}{(q+1)^2} > 0$ for infinitely many $(i,j)'s$ from P. This contradicts that $\mathcal{N} - \lim_j z_{s_j t_j} = 0$. Therefore, the supposition $I^*_{2(\mathcal{N})} - \lim_{st} z_{st} = 0$ is incorrect. As a result, the converse of this theorem has not to be true.

Remark 9.2

Example 9.2 denotes that $I^*_{2(\mathcal{N})}$ – convergence gives $I_{2(\mathcal{N})}$ – convergence but not conversely. We discuss under which condition the converse can hold. Under the property (AP) the converse holds.

Theorem 9.6

Let Ω be a NNS and take I_2 with AP condition. If $I_{2(\mathcal{N})} - \lim z_{jk} = \rho$, then $I^*_{2(\mathcal{N})} - \lim z = \rho$.

Proof: Presume that I_2 satisfies condition (AP) and $I_{2(\mathcal{N})} - \lim z_{jk} = \rho$. Then, for each $\mathfrak{z} > 0, \varepsilon > 0$,

$$\{(j,k) \in \mathbb{N} \times \mathbb{N} : \mathcal{G}(z_{jk} - \rho, \mathfrak{z}) \leq 1 - \varepsilon \text{ or } \mathcal{B}(z_{jk} - \rho, \mathfrak{z}) \geq \varepsilon, \Upsilon(z_{jk} - \rho, \mathfrak{z}) \geq \varepsilon\} \in I_2.$$

We identify the set T_r for $r \in \mathbb{N}$ and $\mathfrak{z} > 0$ as

$$T_r = \left\{(j,k) \in \mathbb{N} \times \mathbb{N} : 1 - \frac{1}{r} \leq \mathcal{G}(z_{jk} - \rho, \mathfrak{z}) < 1 - \frac{1}{r+1} \text{ or } \frac{1}{r+1} < B(z_{jk} - \rho, \mathfrak{z}) \leq \frac{1}{r}, \right.$$
$$\left. \frac{1}{r+1} < Y(z_{jk} - \rho, \mathfrak{z}) \leq \frac{1}{r} \right\}.$$

Clearly $\{T_1, T_2, \ldots\}$ is countable and belongs to I_2, and $T_i \bigcap T_j = \varnothing$ for $i \neq j$. From the condition (AP), there is countable family of sets $\{Q_1, Q_2, \ldots\} \in I_2$ such that $T_i \Delta Q_i$ is a finite set for $i \in \mathbb{N}$ and $Q = \bigcup_{i=1}^{\infty} Q_i \in I_2$. Utilizing the definition of $F(I_2)$ there is a set $V \in F(I_2)$ such that $V = (\mathbb{N} \times \mathbb{N}) / Q$. For proof of theorem, we have to denote that the subsequence (z_{jk}) is convergent to ρ with regard to NN $\mathcal{N}$. Let $\tau > 0$ and $\mathfrak{z} > 0$. Select $w \in \mathbb{N}$ such that $\dfrac{1}{w} < \tau$. Then

$$\{(j,k)\in\mathbb{N}\times\mathbb{N}:\mathcal{G}(z_{jk}-\rho,\mathfrak{z})\le 1-\tau \text{ or } \mathcal{B}(z_{jk}-\rho,\mathfrak{z})\ge\tau, \Upsilon(z_{jk}-\rho,\mathfrak{z})\ge\tau\}\subset$$
$$\left\{(j,k)\in\mathbb{N}\times\mathbb{N}:\mathcal{G}(z_{jk}-\rho,\mathfrak{z})\le 1-\frac{1}{w} \text{ or } \mathcal{B}(z_{jk}-\rho,\mathfrak{z})\ge\frac{1}{w},\ \Upsilon(z_{jk}-\rho,\mathfrak{z})\ge\frac{1}{w}\right\}\subset$$
$$\bigcup_{i=1}^{w+1}T_i.$$

Since $T_i\Delta Q_i$, $i=1,2,\ldots,w+1$ are finite, there exists $(j_0,k_0)\in\mathbb{N}\times\mathbb{N}$ such that

$$\left(\bigcup_{i=1}^{w+1}Q_i\right)\cap\{((j,k):j\ge j_0 \text{ and } k\ge k_0)\}=\left(\bigcup_{i=1}^{w+1}T_i\right)\cap\{((j,k):j\ge j_0 \text{ and } k\ge k_0)\}. \tag{9.2}$$

If $j\ge j_0$ and $k\ge k_0$ and $(j,k)\in V$ then $(j,k)\notin\bigcup_{i=1}^{w+1}Q_i$. By equation (9.2) $(j,k)\notin\bigcup_{i=1}^{w+1}T_i$. So, for every $j\ge j_0$ and $k\ge k_0$ and $(j,k)\in V$, we obtain $\mathcal{G}(z_{jk}-\rho,\mathfrak{z})>1-\tau$ and $\mathcal{B}(z_{jk}-\rho,\mathfrak{z})<\tau, \Upsilon(z_{jk}-\rho,\mathfrak{z})<\tau$. Since $\tau>0$ is arbitrary, we have then $I^*_{2(\mathcal{N})}-\lim z=\rho$.

Theorem 9.7

Following situations are equivalent:

1. $I^*_{2(\mathcal{N})}-\lim z=\rho$.
2. There are two sequences $x=(x_{jk})$ and $y=(y_{jk})$ in Ω such that $z=x+y, \mathcal{N}-\lim x_{jk}=\rho$ and the set $\{((j,k):y_{jk}\neq\theta)\}\in I_2$, where θ indicates the zero element of Ω.

Proof: Let situation (1) hold.

So, there exists a set $\psi=\{(j_m,k_m):\{j_1<j_2<\ldots;k_1<k_2<\ldots\}\}$ of $\mathbb{N}\times\mathbb{N}$ such that

$$\psi\in F(I_2) \text{ and } \mathcal{N}-\lim_m z_{j_mk_m}=\rho. \tag{9.3}$$

We take the sequences (x_{jk}) and (y_{jk}) as follows:

$$x_{jk}=\begin{cases} z_{jk}, & \text{if } (j,k)\in\Psi \\ \rho, & \text{if } (j,k)\in\Psi^c; \end{cases}$$

and $y_{jk} = z_{jk} - x_{jk}$ for $(j,k) \in \mathbb{N} \times \mathbb{N}$. For given $\mathfrak{z} > 0, \varepsilon > 0$ and $(j,k) \in \Psi^c$, we get

$$\mathcal{G}(z_{jk} - \rho, \mathfrak{z}) = 1 > 1 - \varepsilon \text{ and } \mathcal{B}(z_{jk} - \rho, \mathfrak{z}) = 0 < \varepsilon, \Upsilon(z_{jk} - \rho, \mathfrak{z}) = 0 < \varepsilon.$$

Using equation (9.3), we get $\mathcal{N} - \lim x_{jk} = \rho$. Since $\{((j,k) : y_{jk} \neq \theta)\} \subset \Psi^c$, we obtain $\{((j,k) : y_{jk} \neq \theta)\} \in I_2$.

Let the case (2) hold and $\Psi = \{((j,k) : y_{jk} = \theta)\}$. Clearly $\psi \in F(I_2)$ is an infinite set. Let $\psi = \{(j_m, k_m) : \{j_1 < j_2 < \ldots; k_1 < k_2 < \ldots\}\}$.

Since $z_{j_m k_m} = x_{j_m k_m}$ and $\mathcal{N} - \lim_m x_{j_m k_m} = \rho$, $\mathcal{N} - \lim_m z_{j_m k_m} = \rho$. Hence, $I^*_{2(\mathcal{N})} - \lim z = \rho$.

Now, we examine $I_{2(\mathcal{N})}$ and $I^*_{2(\mathcal{N})}$-Cauchy sequences in NNS and demonstrate that $I^*_{2(\mathcal{N})}$-convergence and $I^*_{2(\mathcal{N})}$-Cauchy are equivalent in NNS.

Definition 9.6

A sequence (z_{jk}) is called to be $I_{2(\mathcal{N})}$-Cauchy with regard to NN $\mathcal{N}$ (I_2CaNN) in NNS Ω, if, for every $\mathfrak{z} > 0, \varepsilon > 0$, there exist $T = T(\varepsilon)$ and $R = R(\varepsilon)$ such that, for all, $w \geq T, k, q \geq R$,

$$\{(j,k) \in \mathbb{N} \times \mathbb{N} : \mathcal{G}(z_{jk} - z_{wq}, \mathfrak{z}) \leq 1 - \varepsilon \text{ or } \mathcal{B}(z_{jk} - z_{wq}, \mathfrak{z}) \geq \varepsilon, \Upsilon(z_{jk} - z_{wq}, \mathfrak{z}) \geq \varepsilon\} \in I_2.$$

Definition 9.7

A sequence (z_{jk}) is called to be $I^*_{2(\mathcal{N})}$-Cauchy with regard to NN $\mathcal{N}$ if there is a subset $\Psi = \{(j_m, k_m) : \{j_1 < j_2 < \ldots; k_1 < k_2 < \ldots\}\}$ of $\mathbb{N} \times \mathbb{N}$ such that $\psi \in F(I_2)$ and $(z_{j_m k_m})$ is an ordinary Cauchy sequence with regard to NN $\mathcal{N}$.

The following results are analogues to our Theorems 9.5 and 9.6 that can be proved similarly.

Theorem 9.8

If a sequence (z_{jk}) is $I^*_{2(\mathcal{N})}$-Cauchy with regard to NN $\mathcal{N}$, then it is $I_{2(\mathcal{N})}$-Cauchy with regard to NN $\mathcal{N}$.

Theorem 9.9

Take I_2 with AP condition. If a sequence (z_{jk}) is $I_{2(\mathcal{N})}$-Cauchy with regard to NN $\mathcal{N}$, then it is $I^*_{2(\mathcal{N})}$-Cauchy with regard to NN $\mathcal{N}$.

Theorem 9.10

A sequence (z_{jk}) is $I_{2(\mathcal{N})}$-convergent with regard to NN $\mathcal{N}$ if it is $I_{2(\mathcal{N})}$-Cauchy with regard to NN $\mathcal{N}$

Proof: Necessity: Let $I_{2(\mathcal{N})} - \lim z_{jk} = \rho$. Select $\mathfrak{z} > 0$ such that $(1-\gamma)\circ(1-\gamma) > 1-\varepsilon$ and $\gamma * \gamma < \varepsilon$. For all $\mathfrak{z} > 0$ we consider

$$¥ = \left\{(j,k) \in \mathbb{N}\times\mathbb{N} : \mathcal{G}(z_{jk}-\rho,\mathfrak{z}) \le 1-\gamma \text{ or } \mathcal{B}(z_{jk}-\rho,\mathfrak{z}) \ge \gamma, \Upsilon(z_{jk}-\rho,\mathfrak{z}) \ge \gamma\right\} \in I_2. \tag{9.4}$$

This means that

$$\varnothing \ne ¥^c = \left\{(j,k) \in \mathbb{N}\times\mathbb{N} : \mathcal{G}(z_{jk}-\rho,\mathfrak{z}) > 1-\gamma \text{ and } \mathcal{B}(z_{jk}-\rho,\mathfrak{z}) < \gamma, Y(z_{jk}-\rho,\mathfrak{z}) < \gamma\right\}$$
$$\in F(I_2).$$

Let $(p,q) \in ¥^c$.

So, we obtain $\mathcal{G}(z_{pq}-\rho,\mathfrak{z}) > 1-\gamma$ and $\mathcal{B}(z_{pq}-\rho,\mathfrak{z}) < \gamma, Y(z_{pq}-\rho,\mathfrak{z}) < \gamma$.

Now select

$$Ɣ = \left\{(j,k) \in \mathbb{N}\times\mathbb{N} : \mathcal{G}(z_{jk}-z_{pq},\mathfrak{z}) \le 1-\varepsilon \text{ or } \mathcal{B}(z_{jk}-z_{pq},\mathfrak{z}) \ge \varepsilon, \Upsilon(z_{jk}-z_{pq},\ ) \ge \varepsilon\right\}.$$

We have to prove that $Ɣ \subset ¥$. Let $(j,k) \in Ɣ$. Then, we get

$$\mathcal{G}\left(z_{jk}-z_{pq},\frac{\mathfrak{z}}{2}\right) \le 1-\varepsilon \text{ or } \mathcal{B}\left(z_{jk}-z_{pq},\frac{\mathfrak{z}}{2}\right) \ge \varepsilon, \Upsilon\left(z_{jk}-z_{pq},\frac{\mathfrak{z}}{2}\right) \ge \varepsilon.$$

We have three possible cases. First examine that $\mathcal{G}(z_{jk}-z_{pq},\mathfrak{z}) \le 1-\varepsilon$. Then, we get $\mathcal{G}\left(z_{jk}-\rho,\frac{\mathfrak{z}}{2}\right) \le 1-\gamma$; therefore, $(j,k) \in ¥$. Otherwise, if $\mathcal{G}\left(z_{jk}-\rho,\frac{\mathfrak{z}}{2}\right) > 1-\gamma$, then

$$1-\varepsilon \ge \mathcal{G}(z_{jk}-z_{pq},\mathfrak{z}) \ge \mathcal{G}\left(z_{jk}-\rho,\frac{\mathfrak{z}}{2}\right)\circ\mathcal{G}\left(z_{pq}-\rho,\frac{\mathfrak{z}}{2}\right) > (1-\gamma)\circ(1-\gamma) > 1-\varepsilon,$$

which is impossible. Hence, $Ɣ \subset ¥$.

Consider that $\mathcal{B}(z_{jk}-z_{pq},\mathfrak{z}) \ge \varepsilon$. Then, we get $\mathcal{B}\left(z_{jk}-\rho,\frac{\mathfrak{z}}{2}\right) \ge \gamma$; therefore, $(j,k) \in ¥$. Otherwise, if $\mathcal{B}\left(z_{jk}-\rho,\frac{\mathfrak{z}}{2}\right) \ge \gamma$; then

$$\varepsilon < B(z_{jk}-z_{pq},\mathfrak{z}) \le \mathcal{B}\left(z_{jk}-\rho,\frac{\mathfrak{z}}{2}\right) * \mathcal{B}\left(z_{pq}-\rho,\frac{\mathfrak{z}}{2}\right) < \gamma * \gamma < \varepsilon,$$

which is impossible. Hence, $\mathfrak{Y} \subset \yen$. Similarly for the case $\Upsilon(z_{jk} - z_{pq}, \mathfrak{z}) \geq \varepsilon$, we get $\mathfrak{Y} \subset \yen$.

By equation (9.4), we have $\mathfrak{Y} \in I_2$. So (z_{jk}) is $I_{2(\mathcal{N})}$-Cauchy with regard to NN $\mathcal{N}$.

Sufficiency: Let (z_{jk}) be $I_{2(\mathcal{N})}$-Cauchy with regard to NN $\mathcal{N}$, but not $I_{2(\mathcal{N})}$-convergent. Then, there exist R and Q such that

$$W(\varepsilon, \mathfrak{z}) = \{(j,k) \in \mathbb{N} \times \mathbb{N} : \mathcal{G}(z_{jk} - z_{RQ}, \mathfrak{z}) \leq 1 - \varepsilon \text{ or}$$

$$\mathcal{B}(z_{jk} - z_{RQ}, \mathfrak{z}) \geq \varepsilon, \Upsilon(z_{jk} - z_{RQ}, \mathfrak{z}) \geq \varepsilon\} \in I_2$$

and

$$V(\varepsilon, \mathfrak{z}) = \left\{(j,k) \in \mathbb{N} \times \mathbb{N} : \mathcal{G}\left(z_{jk} - \rho, \frac{\mathfrak{z}}{2}\right) > 1 - \varepsilon \text{ and}\right.$$

$$\left.\mathcal{B}\left(z_{jk} - \rho, \frac{\mathfrak{z}}{2}\right) < \varepsilon, \Upsilon\left(z_{jk} - \rho, \frac{\mathfrak{z}}{2}\right) < \varepsilon\right\} \in I_2$$

equivalently, $V^c(\varepsilon, \mathfrak{z}) \in F(I_2)$. Since

$$\mathcal{G}(z_{jk} - z_{RQ}, \mathfrak{z}) \geq 2\mathcal{G}\left(z_{jk} - \rho, \frac{\mathfrak{z}}{2}\right) > 1 - \varepsilon,$$

and

$$\mathcal{B}(z_{jk} - z_{RQ}, \mathfrak{z}) \leq 2\,\mathcal{B}\left(z_{jk} - \rho, \frac{\mathfrak{z}}{2}\right) < \varepsilon, \qquad \Upsilon(z_{jk} - z_{RQ}, \mathfrak{z}) \leq 2\Upsilon\left(z_{jk} - \rho, \frac{\mathfrak{z}}{2}\right) < \varepsilon.$$

If $\mathcal{G}\left(z_{jk} - \rho, \frac{\mathfrak{z}}{2}\right) > \frac{(1-\varepsilon)}{2}$ and $\mathcal{B}\left(z_{jk} - \rho, \frac{\mathfrak{z}}{2}\right) < \frac{\varepsilon}{2}, \Upsilon\left(z_{jk} - \rho, \frac{\mathfrak{z}}{2}\right) < \frac{\varepsilon}{2}$, respectively, we have $W^c(\varepsilon, \mathfrak{z}) \in I_2$, and so $W(\varepsilon, \mathfrak{z}) \in F(I_2)$, which is a contradiction, as z was $I_{2(\mathcal{N})}$-Cauchy with regard to NN $\mathcal{N}$. Therefore, z has to be $I_{2(\mathcal{N})}$-convergent with regard to NN $\mathcal{N}$.

Likewise, we can give the following:

Theorem 9.11

Choose a NNS Ω. Then each sequence (z_{jk}) is $I^*_{2(\mathcal{N})}$-convergent with regard to NN $\mathcal{N}$ if it is $I^*_{2(\mathcal{N})}$-Cauchy with regard to NN $\mathcal{N}$.

9.3 CONCLUSION

We have examined statistical convergence and ideal convergence of double sequences in NNS. The fundamental characteristic features of NNSs have been studied, and examples are given. Further, statistical completeness, the notions of I_2^*-convergence, I_2-Cauchy and I_2^*-Cauchy for double sequences in NNS are investigated, and notable results are established.

REFERENCES

Atanassov, K. 1986. Intuitionistic fuzzy sets. *Fuzzy Sets and Systems* 20:87–96.

Atanassov, K., G. Pasi, and R. Yager. 2002. Intuitionistic fuzzy interpretations of multi- person multicriteria decision making. *Proceedings of First International IEEE Symposium Intelligent Systems* 1:115–119.

Bera, T., and N.K. Mahapatra. 2017. Neutrosophic soft linear spaces. *Fuzzy Information and Engineering* 9:299–324.

Bera, T., and N.K. Mahapatra. 2018. Neutrosophic soft normed linear spaces. *Neutrosophic Sets and Systems* 23:52–71.

Das, P., and S.K. Ghosal. 2010. Some further results on *I*-Cauchy sequences and condition (AP). *Computers & Mathematics with Applications* 59:2597–2600.

Das, P., E. Savaş, and S.K. Ghosal. 2011. On generalizations of certain summability methods using ideals. *Applied Mathematics Letters* 24:1509–1614.

Debnath, P. 2012. Lacunary ideal convergence in intuitionistic fuzzy normed linear spaces. *Computers & Mathematics with Applications* 63:708–715.

Debnath, P. 2015. Results on lacunary difference ideal convergence in intuitionistic fuzzy normed linear spaces. *Journal of Intelligent & Fuzzy Systems* 28(3):1299–1306.

Debnath, P. 2016. A generalized statistical convergence in intuitionistic fuzzy n-normed linear spaces. *Annals of Fuzzy Mathematics and Informatics* 12(4):559–572.

Debnath, P., and M. Sen. 2014a. Some completeness results in terms of infinite series and quotient spaces in intuitionistic fuzzy *n*-normed linear spaces. *Journal of Intelligent & Fuzzy Systems* 26:975–782.

Debnath, P., and M. Sen. 2014b. Some results of calculus for functions having values in an intuitionistic fuzzy *n*-normed linear space. *Journal of Intelligent & Fuzzy Systems* 26:2983–2991.

Dems, K. 2004. On I-Cauchy sequences. *Real Analysis Exchange* 30:123–128.

Fast, H. 1951. Sur la convergence statistique. *Colloquium Mathematicum* 2:241–244.

George, A., and P. Veeramani. 1994. On some results in fuzzy metric spaces. *Fuzzy Sets and Systems* 64:395–399.

Gürdal, M. 2012. On ideal convergent sequences in 2-normed spaces. *Thai Journal of Mathematics* 4(1):85–91.

Karakuş, S., K. Demirci, and O. Duman. 2008. Statistical convergence on intuitionistic fuzzy normed spaces. *Chaos, Solitons & Fractals* 35(4):763–769.

Kirişci, M., and N. Şimşek. 2020a. Neutrosophic metric spaces. *Mathematical Sciences* 14:241–248.

Kirişci, M., and N. Şimşek. 2020b. Neutrosophic normed spaces and statistical convergence. *The Journal of Analysis*. Doi: 10.1007/s41478-020-00234-00123456789.

Kirişci, M., N. Şimşek, and M. Akyiğit. 2020. Fixed point results for a new metric space. *Mathematical Methods in the Applied Sciences* 2020: 1–7.

Konwar, N., and P. Debnath. 2017. Continuity and Banach contraction principle in intuitionistic fuzzy *n*-normed linear spaces. *Journal of Intelligent & Fuzzy Systems* 33(4):2363–2373.

Konwar, N., B. Davvaz, and P. Debnath. 2018. Approximation of new bounded operators in intuitionistic fuzzy *n*-Banach spaces. *Journal of Intelligent & Fuzzy Systems* 35(6):6301–6312.

Kostyrko, P., M. Macaj, T. Šalát, and M. Sleziak. 2005. I-convergence and extremal I-limit points. *Mathematica Slovaca* 55: 443–464.

Kostyrko, P., T. Šalát, and W. Wilczynki. 2000. I-convergence. *Real Analysis Exchange* 26:669–685.

Kramosil, I., and J. Michalek. 1975. Fuzzy metric and statistical metric spaces. *Kybernetika* 11:336–344.

Mohiuddine, S.A. 2009. Stability of Jensen functional equation in intuitionistic fuzzy normed space. *Chaos, Solitons & Fractals* 42(5):2989–2996.

Mursaleen, M., and S.A. Mohiuddine. 2009a. On lacunary statistical convergence with respect to the intuitionistic fuzzy normed space. *Journal of Computational and Applied Mathematics* 233(2):142–149.

Mursaleen, M., and S.A. Mohiuddine. 2009b. Statistical convergence of double sequences in intuitionistic fuzzy normed spaces. *Chaos, Solitons & Fractals* 41:2414–2421.

Mursaleen, M., and S.A. Mohiuddine. 2009c. On stability of a cubic functional equation in intuitionistic fuzzy normed spaces. *Chaos, Solitons & Fractals* 42(5):2997–3005.

Mursaleen, M., S.A. Mohiuddine, and O.H.H Edely. 2010. On the ideal convergence of double sequences in intuitionistic fuzzy normed spaces. *Computers & Mathematics with Applications* 59:603–611.

Nabiev, A.A., S. Pehlivan, and M. Gürdal. 2007. On *I*-Cauchy sequences. *Taiwanese Journal of Mathematics* 11(2):569–566.

Nabiev, A.A., E. Savaş and M. Gürdal. 2020. *I*-localized sequences in metric spaces. *Facta Universitatis, Series: Mathematics and Informatics* 35(2):459–469.

Park, J.H. 2004. Intuitionistic fuzzy metric spaces. *Chaos, Solitons & Fractals* 22:1039–1046.

Saadati, R., and J.H. Park. 2006. On the intuitionistic fuzzy topological spaces. *Chaos, Solitons & Fractals* 27:331–344.

Savaş, E., and P. Das. 2011. A generalized statistical convergence via ideals. *Applied Mathematics Letters* 24:826–830.

Savaş, E., and M. Gürdal. 2014. Certain summability methods in intuitionistic fuzzy normed spaces. *Journal of Intelligent & Fuzzy Systems* 27(4):1621–1629.

Savaş, E., and M. Gürdal. 2015. A generalized statistical convergence in intuitionistic fuzzy normed spaces. *Science Asia* 41:289–294.

Şimşek, N., and M. Kirişci. 2019. Fixed point theorems in neutrosophic metric spaces. *Sigma Journal of Engineering and Natural Sciences* 10(2):221–230.

Smarandache, F. 1998. *Neutrosophy. Neutrosophic Probability, Set, and Logic. Pro Quest Information & Learning*. American Research Press, Ann Arbor, MI.

Wang, H., F. Smarandache, Y.Q. Zhang, and R. Sunderraman. 2010. Single valued neutrosophic sets. *Multispace&Multistructure* 4:410–413.

Yager, R.R. 2013. Pythagorean fuzzy subsets. In: *Proceedings of the Joint IFSA World Congress and NAFIPS Annual Meeting*, Edmonton, Canada.

Yamancı, U. and M. Gürdal. 2014a. *I*-statistical convergence in 2-normed space. *Arab Journal of Mathematical Sciences* 20(1):41–47.

Yamancı, U. and M. Gürdal. 2014b. *I*-statistically pre Cauchy double sequences. *Global Journal of Mathematical Analysis* 2(4):297–303.

Ye, J. 2014. A multicriteria decision-making method using aggregation operators for simplifedneutrosophic sets. *Journal of Intelligent & Fuzzy Systems* 26:2459–2466.

Zadeh, L.A. 1965. Fuzzy sets. *Information Computing* 8(3):338–353.

10 Intuitionistic Fuzzy Generalized Lucas Ideal Convergent Sequence Spaces Associated with Orlicz Function

Kuldip Raj and Kavita Saini
Shri Mata Vaishno Devi University

CONTENTS

10.1 INTRODUCTION AND PRELIMINARIES

Throughout the chapter, we denote the sequence spaces of real or complex numbers by ω. By c, c_0 and ℓ_∞, we denote the convergent, null and bounded sequence spaces, respectively.

The sequence $\{L_n\}_{n=0}^{\infty}$ of Lucas numbers given by the Fibonacci recurrence relation with different initial conditions is defined by

$$L_0 = 2, L_1 = 1 \quad \text{and} \quad L_n = L_{n-1} + L_{n-2}, \;\; n \geq 2.$$

The ratio of the successive Lucas numbers is known as Golden ratio. Lucas numbers have many interesting properties and applications (Koshy 2001) and (Steven 2008). Also, some basic properties of Lucas numbers are given as follows:

$$\lim_{n\to\infty} \frac{L_n}{L_{n-1}} = \frac{1+\sqrt{5}}{2} = \varphi \qquad \text{(Golden ratio)}$$

$$L_n = \left(\frac{1+\sqrt{5}}{2}\right)^n + \left(\frac{1-\sqrt{5}}{2}\right)^n \qquad \text{(Binet's formula for lucas numbers)}$$

$$L_n^2 - L_{n-1}L_{n+1} = 5(-1)^n \text{ and } \sum_{k=1}^{n} L_k^2 = L_nL_{n+1} - 2 \quad \text{(Additional identities)}.$$

To know more about Lucas and Fibonacci sequences, one may refer to the following articles, see (Demiriz and Ellidokuzoglu 2018), (Raj et al. 2015) and (Yaying et al. 2020).

The generalized Lucas band matrix $\tilde{L} = \left(L_{nk}(\mu,\nu)\right)$ (Karakas and Karakas 2018) is defined by

$$L_{nk}(\mu,\nu) = \begin{cases} \nu\dfrac{L_n}{L_{n-1}}, & k = n-1, \\ \mu\dfrac{L_{n-1}}{L_n}, & k = n, \\ 0, & k > n. \end{cases}$$

The $\tilde{L}$ – transform of a sequence $y = (y_n)$ by

$$z_n = L_{nk}(\mu,\nu)(y) = \mu\frac{L_{n-1}}{L_n}y_n + \nu\frac{L_n}{L_{n-1}}y_{n-1}.$$

Also, Karakas and Karakas (2018) studied the following sequence spaces:

$$\ell_p(\tilde{L}) = \left\{ y = (y_k) \in \omega : \sum_n \left| \mu\frac{L_{n-1}}{L_n}y_n + \nu\frac{L_n}{L_{n-1}}y_{n-1} \right|^p < \infty \right\}, \ 1 \le p < \infty$$

and

$$\ell_\infty(\tilde{L}) = \left\{ y = (y_k) \in \omega : \sup_n \left| \mu\frac{L_{n-1}}{L_n}y_n + \nu\frac{L_n}{L_{n-1}}y_{n-1} \right| < \infty \right\}.$$

The idea of fuzzy number and its arithmetic operation was introduced by Zadeh (1965). Later on, several other authors discussed various applications of the fuzzy sets such as fuzzy ordering, fuzzy measure of fuzzy event, fuzzy mathematical programming, etc. Since then, several mathematicians investigated fuzzy topology to define fuzzy metric space. Park (2004) introduced the notion of intuitionistic fuzzy (shortly, IF) metric space with the help of IF-set (Atanassov 1986). Saadati and Park (2006) introduced intuitionistic fuzzy normed space (IFNS), and then Mursaleen and Lohani (2009) extended this notion to 2-normed space. Lael and Nourouzi (2008) modified the notion of IFNS in order to obtain different topologies from the topology generated by fuzzy norm. To know more about related concepts, one can refer to the following articles: (Hazarika et al. 2020, Kadak et al. 2019, Karakaya et al. 2012, Karakuş et al. 2008), (Mursaleen et al. 2010, Mohiuddine and Lohani 2009, Mursaleen and Mohiuddine 2009b, Khan et al. 2016).

The hypothesis of statistical convergence was first presented by Fast (1951) in 1951 which brings the attention of many researchers from different fields of mathematics. Connor (1988), Fridy (1985) and Salat (1980) linked the notion of statistical convergence with that of summability theory. In Kostyrko et al. (2000) introduced the ideal convergence which is a generalization of statistical convergence. Recently, some researchers obtained an application of these convergence methods in approximation theorems, see (Kadak and Mohiuddine 2018), (Belen and Mohiuddine 2013), (Braha et al. 2014), (Mohiuddine, Asiri and Hazarika 2019), (Mohiuddine, Hazarika and Alghamdi 2019). Consider a nonempty set Z, then a family of sets $I \subset 2^X$ is called ideal if the following conditions are satisfied: (i) $\varnothing \in I$, (ii) for each $Z_1, Z_2 \in I$, we have $Z_1 \cup Z_2 \in I$ and (iii) for each $Z_1 \in I$ and each $Z_2 \subset Z_1$, we have $Z_2 \in I$. A nonempty family of sets $F \subset 2^X$ is a filter on Z if $\phi \notin U$, for each $Z_1, Z_2 \in U$, we have $Z_1 \cap Z_2 \in F$ and each $Z_1 \in F$ and each $Z_1 \subset Z_2$, we h.ave $Z_2 \in F$. For each ideal I there is a filter $F(I)$ corresponding to I, i., $F(I) = \{X \subseteq Z : X^c \in I\}$.

Definition 10.1

Let $K \subseteq \mathbb{N}$, and let $K_n = \{k \leq n : k \in \mathbb{N}\}$. Then, the natural density of K is given by $d(K) = \lim_{n \to \infty} n^{-1} |K_n|$ if the limit exists, where $|K_n|$ denotes the cardinality of the set K_n. A sequence $y = (y_k) \in \omega$ is statistically convergent (Fast 1951) to $l \in \mathbb{R}$ if for every $\varepsilon > 0$, the set $\{k \in \mathbb{N} : |y_k - l| \geq \varepsilon\}$ has natural density zero. We write $st - \lim y = l$.

Definition 10.2

A sequence $y = (y_k) \in \omega$ is statistically Cauchy sequence (Fridy 1985) if for every $\varepsilon > 0$, there exists a number $G = G(\varepsilon)$ such that

$$\lim_n \frac{1}{n} \left| \{k \leq n : |y_k - y_G| \geq \varepsilon\} \right|.$$

Definition 10.3

A sequence $y = (y_k) \in \omega$ is said to be $I -$ convergent (Kostyrko et al. 2000) to $l \in \mathbb{R}$ if for every $\varepsilon > 0$, the set $\{k \in \mathbb{N} : |y_k - l| \geq \varepsilon\} \in I$. We write $I - \lim y = l$.

Definition 10.4

A sequence $y = (y_k) \in \omega$ is said to be $I -$ Cauchy sequence (Dems 2004) if for every $\varepsilon > 0$, there exists a number $G = G(\varepsilon)$ such that

$$\{k \in \mathbb{N} : |y_k - y_G| \geq \varepsilon\} \in I.$$

The Orlicz function $\mathcal{M}:[0,\infty)\to[0,\infty)$ is convex, continuous and nondecreasing function satisfying $\mathcal{M}(0)=0$, $\mathcal{M}(y)>0$ for $y>0$ and $\mathcal{M}(y)\to\infty$ as $y\to\infty$. Lindenstrauss and Tzafriri (1971) defined the Orlicz sequence space $\ell_{\mathcal{M}}$ by

$$\ell_{\mathcal{M}}=\left\{y\in\omega:\sum_{k=1}^{\infty}\mathcal{M}\left(\frac{|y_k|}{\rho}\right)\langle\infty \quad \text{for some } \rho\rangle 0\right\}.$$

The sequence $\mathcal{M}=(\mathcal{M}_k)$ of Orlicz functions is called Musielak-Orlicz function. To know more about sequence spaces, see (Mohiuddine and Hazarika 2017, Mohiuddine et al. 2020, Raj et al. 2018a, b, Tripathy and Hazarika 2011) and references therein.

An increasing nonnegative integer sequence $\theta=(g_{\hat{r}})$ with $g_0=0$ and $g_{\hat{r}}-g_{\hat{r}-1}\to\infty$ as $\hat{r}\to\infty$ is known as lacunary sequence. The intervals determined by θ will be denoted by $I_{\hat{r}}=(g_{\hat{r}-1},g_{\hat{r}}]$. We write $h_{\hat{r}}=g_{\hat{r}}-g_{\hat{r}-1}$ and $q_{\hat{r}}$ denotes the ratio $\frac{g_{\hat{r}}}{g_{\hat{r}-1}}$. The space of lacunary strongly convergent sequence was defined by Freedman et al. (1978) as follows:

$$R_\theta=\left\{y=(y_k):\lim_{\hat{r}\to\infty}\frac{1}{h_{\hat{r}}}\sum_{k\in I_{\hat{r}}}|y_k-L|=0 \quad \text{for some } L\right\}.$$

The space R_θ is a $BK-$ space with the norm

$$\| y \|=\sup\left(\frac{1}{h_{\hat{r}}}\sum_{k\in I_{\hat{r}}}|y_k|\right).$$

By using lacunary sequence, Fridy and Orhan (1993) introduced the notion of lacunary statistical convergence. To know more about lacunary sequence and related concepts, one can refer to (Mohiuddine and Alamri 2019, Mursaleen and Mohiuddine 2009a, Tripathy et al. 2012, Yaying and Hazarika 2019, 2020) and references therein.

Now, we recall some definitions which will be useful in this chapter.

Definition 10.5

A binary operation $*:[0,1]\times[0,1]\to[0,1]$ is continuous $t-$ norm (Schweizer and Sklar 1960) if it satisfies the following conditions:

a. $*$ is commutative and associative;
b. $*$ is continuous;
c. $a*1=a$ for all $a\in[0,1]$;
d. $a*b\le c*d$ whenever $a\le c$ and $b\le d$ and $a,b,c,d\in[0,1]$.

Definition 10.6

A binary operation $\lozenge : [0,1]\times[0,1]\rightarrow[0,1]$ is continuous $t-$ conorm (Schweizer and Sklar 1960) if it satisfies the following conditions:

a. $\lozenge$ is commutative and associative;
b. $\lozenge$ is continuous;
c. $a\lozenge 0 = a$ for all $a \in [0,1]$;
d. $a\lozenge b \le c\lozenge d$ whenever $a \le c$ and $b \le d$ and $a,b,c,d \in [0,1]$.

Definition 10.7

The 5-tuple $(U,\eta,\vartheta,*,\lozenge)$ is said to be an intuitionistic fuzzy normed space (IFNS) (Lael and Nourouzi 2008) if U is a real vector space, $*$ is a continuous $t-$ norm, $\lozenge$ is a continuous $t-$ conorm and η,ϑ are fuzzy sets on $U\times\mathbb{R}$ satisfying the following conditions for every $y,z \in U$ and $m,n \in \mathbb{R}$:

a. $\eta(y,m) = 0$ for all non-positive real number m,
b. $\eta(y,m) = 1$ for all $m \in \mathbb{R}^+$ if $y = 0$,
c. $\eta(cy,m) = \eta\left(y,\frac{m}{|c|}\right)$ for all $m \in \mathbb{R}^+$ and $c \ne 0$,
d. $\eta(y+z,m+n) \ge \eta(y,m)*\eta(z,n)$,
e. $\lim_{m\to\infty}\eta(y,m) = 1$ and $\lim_{m\to 0}\eta(y,m) = 0$,
f. $\vartheta(y,m) = 1$ for all nonpositive real number m,
g. $\vartheta(y,m) = 0$ for all $m \in \mathbb{R}^+$ if $y = 0$,
h. $\vartheta(cy,m) = \vartheta\left(y,\frac{m}{|c|}\right)$ for all $m \in \mathbb{R}^+$ and $c \ne 0$,
i. $\vartheta(y+z,m+n) \le \vartheta(y,m)\lozenge\vartheta(z,n)$,
j. $\lim_{m\to\infty}\vartheta(y,m) = 0$ and $\lim_{m\to 0}\vartheta(y,m) = 1$.

We call (η,ϑ) an intuitionistic fuzzy norm (IFN). It is easy to see that for every $y \in U$, the functions $\eta(y,.)$ and $\vartheta(y,.)$ are nondecreasing and nonincreasing on $\mathbb{R}$, respectively.

Definition 10.8

Let $(U,\eta,\vartheta,*,\lozenge)$ be an IFNS. A sequence $y = (y_k)$ is convergent to $l \in U$ with respect to the intuitionistic fuzzy norm (η,ϑ) if for every $\varepsilon,m > 0$, there exists a number $G \in \mathbb{N}$ with $\eta(y_k - l,m) > 1-\varepsilon$ and $\vartheta(y_k - l,m) < \varepsilon$ for all $k \ge G$. We write $(\eta,\vartheta)-\lim y = l$.

Definition 10.9

A sequence $y = (y_k)$ is said to be Cauchy sequence with respect to the intuitionistic fuzzy norm (η, ϑ) if for each $\varepsilon, m > 0$, there exists a number $G \in \mathbb{N}$ with $\eta(y_k - y_n, m) > 1 - \varepsilon$ and $\vartheta(y_k - y_n, m) < \varepsilon$ for all $k, n \geq G$.

Definition 10.10

Let $(U, \eta, \vartheta, *, \lozenge)$ be an IFNS. A sequence $y = (y_k)$ is said to be $I_\theta -$ convergent to $l \in U$ with respect to the intuitionistic fuzzy norm (η, ϑ) if for each $\varepsilon, m > 0$, we have

$$\left\{\hat{r} \in \mathbb{N} : \frac{1}{h_{\hat{r}}} \sum_{k \in I_{\hat{r}}} \eta(y_k - l, m) \leq 1 - \varepsilon \ or \ \vartheta(y_k - l, m) \geq \varepsilon \right\} \in I$$

and it is denoted by $I_\theta^{(\eta,\vartheta)} - \lim y = l$.

Definition 10.11

A sequence $y = (y_k)$ is said to be generalized Lucas $I -$ convergent to $l \in \mathbb{R}$, if for every $\varepsilon > 0$, the set

$$\left\{k \in \mathbb{N} : \left|y_k^{'} - l\right| \geq \varepsilon\right\} \in I,$$

where $y_k^{'} = \tilde{L} y_k$.

Definition 10.12

A sequence $y = (y_k)$ is said to be generalized Lucas $I -$ Cauchy sequence if for every $\varepsilon > 0$, there exists a number $G = G(\varepsilon)$ such that

$$\left\{k \in \mathbb{N} : \left|y_k^{'} - y_G^{'}\right| \geq \varepsilon\right\} \in I.$$

10.2 INTUITIONISTIC GENERALIZED LUCAS $I -$ CONVERGENT SEQUENCE SPACES

In this section, we introduce the intuitionistic generalized Lucas $I -$ convergent sequence spaces defined by Orlicz function and lacunary sequence as follows:

$$\tilde{L}_{(\eta,\vartheta)}^{I,\theta}(\mathcal{M}) = \{y = (y_k) \in \ell_\infty : \{\hat{r} \in \mathbb{N} : \frac{1}{h_{\hat{r}}} \sum_{k \in I_{\hat{r}}} \mathcal{M}\left(\frac{\eta(y_k^{'} - l, m)}{\rho}\right) \leq 1 - \varepsilon$$

$$\text{or} \frac{1}{h_{\hat{r}}} \sum_{k \in I_{\hat{r}}} \mathcal{M}\left(\frac{\vartheta\left(y_k^{'} - l, m\right)}{\rho}\right) \geq \varepsilon\} \in I\}$$

and

$$\tilde{L}_{0(\eta,\vartheta)}^{I,\theta}(\mathcal{M}) = \{y = (y_k) \in \ell_\infty : \{\hat{r} \in \mathbb{N} : \frac{1}{h_{\hat{r}}} \sum_{k \in I_{\hat{r}}} \mathcal{M}\left(\frac{\eta\left(y_k^{'}, m\right)}{\rho}\right) \leq 1 - \varepsilon$$

$$\text{or } \frac{1}{h_{\hat{r}}} \sum_{k \in I_{\hat{r}}} \mathcal{M}\left(\frac{\vartheta\left(y_k^{'}, m\right)}{\rho}\right) \geq \varepsilon\} \in I\}.$$

We define an open ball with center y and radius $\tilde{r}$ with respect to m by

$$B(y, \tilde{r}, m)(\tilde{L}, \mathcal{M}) = \left\{z = (z_k) \in \ell_\infty : \left\{\hat{r} \in \mathbb{N} : \frac{1}{h_{\hat{r}}} \sum_{k \in I_{\hat{r}}} \mathcal{M}\left(\frac{\eta\left(y_k^{'} - z_k^{'}, m\right)}{\rho}\right) > 1 - \hat{r}\right.\right.$$

$$\left.\left.\text{and } \frac{1}{h_{\hat{r}}} \sum_{k \in I_{\hat{r}}} \mathcal{M}\left(\frac{\vartheta\left(y_k^{'} - z_k^{'}, m\right)}{\rho}\right) < \tilde{r}\right\}\right\}$$

Theorem 10.1

The spaces $\tilde{L}_{(\eta,\vartheta)}^{I,\theta}(\mathcal{M})$ and $\tilde{L}_{0(\eta,\vartheta)}^{I,\theta}(\mathcal{M})$ are linear spaces over $\mathbb{R}$.

Proof: We will prove the result for $\tilde{L}_{(\eta,\vartheta)}^{I,\theta}(\mathcal{M})$. The proof for the other case will follow the same pattern. Let $y = (y_k), z = (z_k) \in \tilde{L}_{(\eta,\vartheta)}^{I,\theta}(\mathcal{M})$ and α, β be scalars. Since $\mathcal{M}$ is an Orlicz function, and $\theta = (g_{\hat{r}})$ is an increasing sequence of non-negative integers, we have

$$\frac{1}{h_{\hat{r}}} \sum_{k \in I_{\hat{r}}} \mathcal{M}\left(\frac{1 - \varepsilon}{\rho_3}\right) > 1 - \varepsilon \text{ and } \frac{1}{h_{\hat{r}}} \sum_{k \in I_{\hat{r}}} \mathcal{M}\left(\frac{\varepsilon}{\rho_3}\right) < \varepsilon.$$

Therefore, we have

$$B_1 = \left\{\hat{r} \in \mathbb{N} : \frac{1}{h_{\hat{r}}} \sum_{k \in I_{\hat{r}}} \mathcal{M}\left(\frac{\eta\left(y_k^{'} - l_1, \frac{m}{2|\alpha|}\right)}{\rho_1}\right) \leq 1 - \varepsilon\right.$$

$$\left.\text{or } \frac{1}{h_{\hat{r}}} \sum_{k \in I_{\hat{r}}} \mathcal{M}\left(\frac{\vartheta\left(y_k^{'} - l_1, \frac{m}{2|\alpha|}\right)}{\rho_1}\right)\right\} \geq \varepsilon \in I$$

$$B_2 = \left\{ \hat{r} \in \mathbb{N} : \frac{1}{h_{\hat{r}}} \sum_{k \in I_{\hat{r}}} \mathcal{M}\left(\frac{\eta\left(z_k^{'} - l_2, \frac{m}{2|\beta|} \right)}{\rho_2} \right) \leq 1 - \varepsilon \right.$$

$$\left. \text{or } \frac{1}{h_{\hat{r}}} \sum_{k \in I_{\hat{r}}} \mathcal{M}\left(\frac{\vartheta\left(z_k^{'} - l_2, \frac{m}{2|\beta|} \right)}{\rho_2} \right) \right\} \geq \varepsilon \in I.$$

Thus,

$$B_1^c = \left\{ \hat{r} \in \mathbb{N} : \frac{1}{h_{\hat{r}}} \sum_{k \in I_{\hat{r}}} \mathcal{M}\left(\frac{\eta\left(y_k^{'} - l_1, \frac{m}{2|\alpha|} \right)}{\rho_1} \right) > 1 - \varepsilon \right.$$

$$\left. \text{or } \frac{1}{h_{\hat{r}}} \sum_{k \in I_{\hat{r}}} \mathcal{M}\left(\frac{\vartheta\left(y_k^{'} - l_1, \frac{m}{2|\alpha|} \right)}{\rho_1} \right) \right\} < \varepsilon \in F(I)$$

$$B_2^c = \left\{ \hat{r} \in \mathbb{N} : \frac{1}{h_{\hat{r}}} \sum_{k \in I_{\hat{r}}} \mathcal{M}\left(\frac{\eta\left(z_k^{'} - l_1, \frac{m}{2|\beta|} \right)}{\rho_2} \right) > 1 - \varepsilon \right.$$

$$\left. \text{or } \frac{1}{h_{\hat{r}}} \sum_{k \in I_{\hat{r}}} \mathcal{M}\left(\frac{\vartheta\left(z_k^{'} - l_1, \frac{m}{2|\beta|} \right)}{\rho_2} \right) \right\} < \varepsilon \in F(I)$$

Define the set $B_3 = B_1 \bigcup B_2$, so that $B_3 \in I$. It follows that B_3^c is a non-empty set in $F(I)$.

Suppose $\rho_3 = \max\left\{ 2|\alpha|\rho_1, 2|\beta|\rho_2 \right\}$ and we will show that

$$B_3^c \subset \{ \hat{r} \in \mathbb{N} : \frac{1}{h_{\hat{r}}} \sum_{k \in I_{\hat{r}}} \mathcal{M}\left(\frac{\eta\left(\left(\alpha y_k^{'} + \beta z_k^{'} \right) - \left(\alpha l_1 + \beta l_2 \right), m \right)}{\rho_3} \right) > 1 - \varepsilon$$

$$\text{or}\ \frac{1}{h_{\hat{r}}}\sum_{k\in I_{\hat{r}}}\mathcal{M}\left(\frac{\vartheta\left(\left(\alpha y_k' + \beta z_k'\right) - \left(\alpha l_1 + \beta l_2\right), m\right)}{\rho_3}\right) < \varepsilon.$$

Since $\mathcal{M}$ is non-decreasing and convex function, we have

$$\frac{1}{h_{\hat{r}}}\sum_{k\in I_{\hat{r}}}\mathcal{M}\left(\frac{\eta\left(\left(\alpha y_k' + \beta z_k'\right) - \left(\alpha l_1 + \beta l_2\right), m\right)}{\rho_3}\right)$$

$$\geq \frac{1}{h_{\hat{r}}}\sum_{k\in I_{\hat{r}}}\mathcal{M}\left(\frac{\eta\left(\alpha y_k' - \alpha l_1, \frac{m}{2}\right) * \eta\left(\beta z_k' - \beta l_2, \frac{m}{2}\right)}{\rho_3}\right)$$

$$= \frac{1}{h_{\hat{r}}}\sum_{k\in I_{\hat{r}}}\mathcal{M}\left(\frac{\eta\left(y_k' - l_1, \frac{m}{2|\alpha|}\right) * \eta\left(z_k' - l_2, \frac{m}{2|\beta|}\right)}{\rho_3}\right)$$

$$\geq \frac{1}{h_{\hat{r}}}\sum_{k\in I_{\hat{r}}}\mathcal{M}\left(\frac{(1-\varepsilon)*(1-\varepsilon)}{\rho_3}\right) > \frac{1}{h_{\hat{r}}}\sum_{k\in I_{\hat{r}}}\mathcal{M}\left(\frac{1-\varepsilon}{\rho_3}\right) > 1-\varepsilon$$

and

$$\frac{1}{h_{\hat{r}}}\sum_{k\in I_{\hat{r}}}\mathcal{M}\left(\frac{\vartheta\left(\left(\alpha y_k' + \beta z_k'\right) - \left(\alpha l_1 + \beta l_2\right), m\right)}{\rho_3}\right)$$

$$\leq \frac{1}{h_{\hat{r}}}\sum_{k\in I_{\hat{r}}}\mathcal{M}\left(\frac{\vartheta\left(\alpha y_k' - \alpha l_1, \frac{m}{2}\right) \Diamond \vartheta\left(\beta z_k' - \beta l_2, \frac{m}{2}\right)}{\rho_3}\right)$$

$$= \frac{1}{h_{\hat{r}}}\sum_{k\in I_{\hat{r}}}\mathcal{M}\left(\frac{\vartheta\left(y_k' - l_1, \frac{m}{2|\alpha|}\right) \Diamond \vartheta\left(z_k' - l_2, \frac{m}{2|\beta|}\right)}{\rho_3}\right)$$

$$\leq \frac{1}{h_{\hat{r}}}\sum_{k\in I_{\hat{r}}}\mathcal{M}\left(\frac{\varepsilon \Diamond \varepsilon}{\rho_3}\right) < \frac{1}{h_{\hat{r}}}\sum_{k\in I_{\hat{r}}}\mathcal{M}\left(\frac{\varepsilon}{\rho_3}\right) < \varepsilon.$$

Thus

$$B_3^c \subset \{\hat{r} \in \mathbb{N} : \frac{1}{h_{\hat{r}}} \sum_{k \in I_{\hat{r}}} \mathcal{M}\left(\frac{\eta\left(\left(\alpha y_k^{'} + \beta z_k^{'}\right) - \left(\alpha l_1 + \beta l_2\right), m\right)}{\rho_3}\right) > 1 - \varepsilon$$

$$\text{or} \frac{1}{h_{\hat{r}}} \sum_{k \in I_{\hat{r}}} \mathcal{M}\left(\frac{\vartheta\left(\left(\alpha y_k^{'} + \beta z_k^{'}\right) - \left(\alpha l_1 + \beta l_2\right), m\right)}{\rho_3}\right) < \varepsilon\}$$

Hence, $\tilde{L}^{I,\theta}_{(\eta,\vartheta)}(\mathcal{M})$ is a linear space.

Theorem 10.2

Every open ball $B(y,\tilde{r},m)(\tilde{L},\mathcal{M})$ is an open set in $\tilde{L}^{I,\theta}_{(\eta,\vartheta)}(\mathcal{M})$.

Proof: Let $B(y,\tilde{r},m)(\tilde{L},\mathcal{M})$ be an open ball with center y and radius $\tilde{r}$ with respect to m, that is

$$B(y,\tilde{r},m)(\tilde{L},\mathcal{M}) = \{z = (z_k) \in \ell_\infty : \{\hat{r} \in \mathbb{N} : \frac{1}{h_{\hat{r}}} \sum_{k \in I_{\hat{r}}} \mathcal{M}\left(\frac{\eta\left(y_k^{'} - z_k^{'}, m\right)}{\rho}\right) > 1 - \tilde{r}$$

$$\text{and} \frac{1}{h_{\hat{r}}} \sum_{k \in I_{\hat{r}}} \mathcal{M}\left(\frac{\vartheta\left(y_k^{'} - z_k^{'}, m\right)}{\rho}\right) < \tilde{r}\}\}$$

Let $z_k \in B(y,\tilde{r},m)(\tilde{L},\mathcal{M})$. Then

$$\frac{1}{h_{\hat{r}}} \sum_{k \in I_{\hat{r}}} \mathcal{M}\left(\frac{\eta\left(y_k^{'} - z_k^{'}, m\right)}{\rho}\right) > 1 - \tilde{r}$$

$$\text{and} \frac{1}{h_{\hat{r}}} \sum_{k \in I_{\hat{r}}} \mathcal{M}\left(\frac{\vartheta\left(y_k^{'} - z_k^{'}, m\right)}{\rho}\right) < \tilde{r}.$$

Then, $\exists m_0 \in (0,1)$ such that

$$\frac{1}{h_{\hat{r}}} \sum_{k \in I_{\hat{r}}} \mathcal{M}\left(\frac{\eta\left(y_k^{'} - z_k^{'}, m_0\right)}{\rho}\right) > 1 - \tilde{r}$$

$$\text{and} \frac{1}{h_{\hat{r}}} \sum_{k \in I_{\hat{r}}} \mathcal{M}\left(\frac{\vartheta\left(y_k^{'} - z_k^{'}, m_0\right)}{\rho}\right) < \tilde{r}.$$

Put $s_0 = \frac{1}{h_{\hat{r}}} \sum_{k \in I_{\hat{r}}}, \mathcal{M}\left(\frac{\eta\left(y_k^{'} - z_k^{'}, m_0\right)}{\rho} \right)$ we have $s_0 > 1 - \tilde{r}$, there exists $t \in (0,1)$ such that $s_0 > 1 - t > 1 - \tilde{r}$. For $s_0 > 1 - t$, we have $r_1, r_2 \in (0,1)$ such that $s_0 * r_1 > 1 - t$ and $(1 - s_0) \Diamond (1 - r_2) \le t$. Put $r_3 = \max\{r_1, r_2\}$. Consider the ball $B(z, 1 - r_3, m - m_0)(\tilde{L}, \mathcal{M})$. We prove that $B(z, 1 - r_3, m - m_0)(\tilde{L}, \mathcal{M}) \subset B(y, \tilde{r}, m)(\tilde{L}, \mathcal{M})$. Consider $q = (q_k) \in B(z, 1 - r_3, m - m_0)(\tilde{L}, \mathcal{M})$, then

$$\frac{1}{h_{\hat{r}}} \sum_{k \in I_{\hat{r}}} \mathcal{M}\left(\frac{\eta\left(z_k^{'} - q_k^{'}, m - m_0\right)}{\rho} \right) > r_3$$

$$\text{and} \frac{1}{h_{\hat{r}}} \sum_{k \in I_{\hat{r}}} \mathcal{M}\left(\frac{\vartheta\left(z_k^{'} - q_k^{'}, m - m_0\right)}{\rho} \right) < 1 - r_3.$$

Therefore,

$$\frac{1}{h_{\hat{r}}} \sum_{k \in I_{\hat{r}}} \mathcal{M}\left(\frac{\eta\left(y_k^{'} - q_k^{'}, m\right)}{\rho} \right) \ge \frac{1}{h_{\hat{r}}} \sum_{k \in I_{\hat{r}}} \mathcal{M}\left(\frac{\eta\left(y_k^{'} - z_k^{'}, m_0\right) * \eta\left(z_k^{'} - q_k^{'}, m - m_0\right)}{\rho} \right)$$

$$\ge s_0 * r_3 \ge s_0 * r_1 > 1 - t > 1 - \tilde{r}$$

and

$$\frac{1}{h_{\hat{r}}} \sum_{k \in I_{\hat{r}}} \mathcal{M}\left(\frac{\vartheta\left(y_k^{'} - q_k^{'}, m\right)}{\rho} \right) \le \frac{1}{h_{\hat{r}}} \sum_{k \in I_{\hat{r}}} \mathcal{M}\left(\frac{\vartheta\left(y_k^{'} - z_k^{'}, m_0\right) \Diamond \eta\left(z_k^{'} - q_k^{'}, m - m_0\right)}{\rho} \right)$$

$$\le (1 - s_0) \Diamond (1 - r_3) \le (1 - s_0) \Diamond (1 - r_2) \le t < \tilde{r}.$$

Hence, $q \in B(y, \tilde{r}, m)(\tilde{L}, \mathcal{M})$ and therefore $B(z, 1 - r_3, m - m_0)(\tilde{L}, \mathcal{M}) \subset B(y, \tilde{r}, m)(\tilde{L}, \mathcal{M})$.

Definition 10.13

Let $(U, \eta, \vartheta, *, \Diamond)$ be an IFNS. Define

$\tau_{(\eta,\vartheta)}^{I,\theta}(\tilde{L}, \mathcal{M}) = \{P \subset \tilde{L}_{(\eta,\vartheta)}^{I,\theta}(\mathcal{M})$: for each $y \in P$ there exists $m > 0$ and $r \in (0,1)$ such that

$$B(y,\tilde{r},m)(\tilde{L},\mathcal{M})\subset P\}.$$

Then, $\tau^{I,\theta}_{(\eta,\vartheta)}(\tilde{L},\mathcal{M})$ is a topology on $\tilde{L}^{I,\theta}_{(\eta,\vartheta)}(\mathcal{M})$.

Theorem 10.3

$\tilde{L}^{I,\theta}_{(\eta,\vartheta)}(\mathcal{M})$ and $\tilde{L}^{I,\theta}_{(\eta,\vartheta)}(\mathcal{M})$ are Hausdorff spaces.

Proof: We prove the result for $\tilde{L}^{I,\theta}_{(\eta,\vartheta)}(\mathcal{M})$. Similar result can be established for $\tilde{L}^{I,\theta}_{(\eta,\vartheta)}(\mathcal{M})$. Let $y,z\in\tilde{L}^{I,\theta}_{(\eta,\vartheta)}(\mathcal{M})$ such that $y\neq z$. Then

$$0<\frac{1}{h_{\hat{r}}}\sum_{k\in I_{\hat{r}}}\mathcal{M}\left(\frac{\eta(y_k^{'}-z_k^{'},m)}{\rho}\right)<1$$

$$\text{and } 0<\frac{1}{h_{\hat{r}}}\sum_{k\in I_{\hat{r}}}\mathcal{M}\left(\frac{\vartheta(y_k^{'}-z_k^{'},m)}{\rho}\right)<1.$$

Put $s_1=\frac{1}{h_{\hat{r}}}\sum_{k\in I_{\hat{r}}}\mathcal{M}\left(\frac{\eta(y_k^{'}-z_k^{'},m)}{\rho}\right)$ and $s_2=\frac{1}{h_{\hat{r}}}\sum_{k\in I_{\hat{r}}}\mathcal{M}\left(\frac{\vartheta(y_k^{'}-z_k^{'},m)}{\rho}\right)$ and $r=\max\{s_1,1-s_2\}$. For each $s_0\in(r,1)$, there exist s_3 and s_4 such that $s_3*s_3\geq s_0$ and $(1-s_4)\Diamond(1-s_4)\leq(1-s_0)$. Put $s_5=\max\{s_3,1-s_4\}$ and consider the open balls $B\left(y,1-s_5,\frac{m}{2}\right)(\tilde{L},\mathcal{M})$ and $B\left(z,1-s_5,\frac{m}{2}\right)(\tilde{L},\mathcal{M})$. Clearly, $B\left(y,1-s_5,\frac{m}{2}\right)(\tilde{L},\mathcal{M})\bigcap B\left(z,1-s_5,\frac{m}{2}\right)(\tilde{L},\mathcal{M})=\varnothing$. Suppose on the contrary $q\in B\left(y,1-s_5,\frac{m}{2}\right)(\tilde{L},\mathcal{M})\bigcap B\left(z,1-s_5,\frac{m}{2}\right)(\tilde{L},\mathcal{M})$, we have

$$s_1=\frac{1}{h_{\hat{r}}}\sum_{k\in I_{\hat{r}}}\mathcal{M}\left(\frac{\eta(y_k^{'}-z_k^{'},m)}{\rho}\right)$$

$$\geq\frac{1}{h_{\hat{r}}}\sum_{k\in I_{\hat{r}}}\mathcal{M}\left(\frac{\eta\left(y_k^{'}-q_k^{'},\frac{m}{2}\right)*\eta\left(q_k^{'}-z_k^{'},\frac{m}{2}\right)}{\rho}\right)$$

$$\geq s_5*s_5\geq s_3*s_3\geq s_0>s_1$$

and

$$s_2=\frac{1}{h_{\hat{r}}}\sum_{k\in I_{\hat{r}}}\mathcal{M}\left(\frac{\vartheta(y_k^{'}-z_k^{'},m)}{\rho}\right)$$

$$\leq \frac{1}{h_{\hat{r}}}\sum_{k\in I_{\hat{r}}}\mathcal{M}\left(\frac{\vartheta\left(y_k^{'}-q_k^{'},\frac{m}{2}\right)\Diamond\vartheta\left(q_k^{'}-z_k^{'},\frac{m}{2}\right)}{\rho}\right)$$

$$\leq (1-s_5)\Diamond(1-s_5)\leq(1-s_4)\Diamond(1-s_4)\leq(1-s_0)<s_2$$

which is a contradiction. Thus, $\tilde{L}^{I,\theta}_{(\eta,\vartheta)}(\mathcal{M})$ is a Hausdorff space.

Theorem 10.4

Let $\tilde{L}^{I,\theta}_{(\eta,\vartheta)}(\mathcal{M})$ be an IFNS and $\tau^{I,\theta}_{(\eta,\vartheta)}(\tilde{L},\mathcal{M})$ be a topology on $\tilde{L}^{I,\theta}_{(\eta,\vartheta)}(\mathcal{M})$ Then a sequence $(y_k)\in\tilde{L}^{I,\theta}_{(\eta,\vartheta)}(\mathcal{M})$ converges to l if

$$\frac{1}{h_{\hat{r}}}\sum_{k\in I_{\hat{r}}}\mathcal{M}\left(\frac{\eta\left(y_k^{'}-l,m\right)}{\rho}\right)\to 1$$

$$\text{and } \frac{1}{h_{\hat{r}}}\sum_{k\in I_{\hat{r}}}\mathcal{M}\left(\frac{\vartheta\left(y_k^{'}-l,m\right)}{\rho}\right)\to 0 \quad \text{as k}\to\infty.$$

Proof: Let the sequence $(y_k)\to l$ and $m>0$. Then, for $r\in(0,1),\exists c_0\in\mathbb{N}$ such that $(y_k)\in B(y,\tilde{r},m)(\tilde{L},\mathcal{M})$ for all $k\geq c_0$. Therefore,

$$1-\frac{1}{h_{\hat{r}}}\sum_{k\in I_{\hat{r}}}\mathcal{M}\left(\frac{\eta\left(y_k^{'}-l,m\right)}{\rho}\right)<r$$

$$\text{and } \frac{1}{h_{\hat{r}}}\sum_{k\in I_{\hat{r}}}\mathcal{M}\left(\frac{\vartheta\left(y_k^{'}-l,m\right)}{\rho}\right)<r.$$

Thus,

$$\frac{1}{h_{\hat{r}}}\sum_{k\in I_{\hat{r}}}\mathcal{M}\left(\frac{\eta\left(y_k^{'}-l,m\right)}{\rho}\right)\to 1$$

$$\text{and } \frac{1}{h_{\hat{r}}}\sum_{k\in I_{\hat{r}}}\mathcal{M}\left(\frac{\vartheta\left(y_k^{'}-l,m\right)}{\rho}\right)\to 0 \quad \text{as k}\to\infty.$$

Conversely, for each $m>0$, we assume that

$$\frac{1}{h_{\hat{r}}}\sum_{k\in I_{\hat{r}}}\mathcal{M}\left(\frac{\eta\left(y_k^{'}-l,m\right)}{\rho}\right)\to 1$$

$$\text{and } \frac{1}{h_{\hat{r}}}\sum_{k\in I_{\hat{r}}}\mathcal{M}\left(\frac{\vartheta\left(y_k^{'}-l,m\right)}{\rho}\right)\rightarrow 0 \quad \text{as k} \rightarrow \infty.$$

For $r\in(0,1),\exists c_0\in\mathbb{N}$ such that

$$1-\frac{1}{h_{\hat{r}}}\sum_{k\in I_{\hat{r}}}\mathcal{M}\left(\frac{\eta\left(y_k^{'}-l,m\right)}{\rho}\right)<r$$

$$\text{and } \frac{1}{h_{\hat{r}}}\sum_{k\in I_{\hat{r}}}\mathcal{M}\left(\frac{\vartheta\left(y_k^{'}-l,m\right)}{\rho}\right)<r$$

for all $k\geq c_0$, which implies

$$\frac{1}{h_{\hat{r}}}\sum_{k\in I_{\hat{r}}}\mathcal{M}\left(\frac{\eta\left(y_k^{'}-l,m\right)}{\rho}\right)>1-r$$

$$\text{and } \frac{1}{h_{\hat{r}}}\sum_{k\in I_{\hat{r}}}\mathcal{M}\left(\frac{\vartheta\left(y_k^{'}-l,m\right)}{\rho}\right)<r$$

Thus, $\left(y_k\right)\in\tilde{L}^{I,\theta}_{(\eta,\vartheta)}\left(\mathcal{M}\right)$ for all $k\geq c_0$ and hence, $y_k\rightarrow l$.

Theorem 10.5

A sequence $y=\left(y_k\right)\in\tilde{L}^{I,\theta}_{(\eta,\vartheta)}\left(\mathcal{M}\right)$ is $I-$ convergent if for every $\varepsilon>0$ and $m>0$ there exists a number $G=G\left(y,\varepsilon,m\right)$ such that

$$\{\hat{r}\in\mathbb{N}:\frac{1}{h_{\hat{r}}}\sum_{k\in I_{\hat{r}}}\mathcal{M}\left(\frac{\eta\left(y_k^{'}-y_G^{'},\frac{m}{2}\right)}{\rho}\right)>1-\varepsilon$$

$$\text{and } \frac{1}{h_{\hat{r}}}\sum_{k\in I_{\hat{r}}}\mathcal{M}\left(\frac{\vartheta\left(y_k^{'}-y_G^{'},\frac{m}{2}\right)}{\rho}\right)<\varepsilon\}\in F(I).$$

Proof: Suppose that $I_\theta^{(\eta,\vartheta)}-\lim y=l$ and let $\varepsilon>0,t>0$. For a given $\varepsilon>0$, choose $n>0$ such that $(1-\varepsilon)*(1-\varepsilon)>1-n$ and $\varepsilon\Diamond\varepsilon<n$. Then for each $y\in\tilde{L}^{I,\theta}_{(\eta,\vartheta)}\left(\mathcal{M}\right)$,

$$B=\{\hat{r}\in\mathbb{N}:\frac{1}{h_{\hat{r}}}\sum_{k\in I_{\hat{r}}}\mathcal{M}\left(\frac{\eta\left(y_k^{'}-y_G^{'},\frac{m}{2}\right)}{\rho}\right)\leq 1-\varepsilon$$

$$\text{or} \frac{1}{h_{\hat{r}}} \sum_{k \in I_{\hat{r}}} \mathcal{M}\left(\frac{\vartheta\left(y_k' - y_G', \frac{m}{2} \right)}{\rho} \right) \geq \varepsilon\} \in I$$

which implies that

$$B^c = \{\hat{r} \in \mathbb{N} : \frac{1}{h_{\hat{r}}} \sum_{k \in I_{\hat{r}}} \mathcal{M}\left(\frac{\eta\left(y_k' - y_G', \frac{m}{2} \right)}{\rho} \right) > 1 - \varepsilon$$

$$\text{or} \frac{1}{h_{\hat{r}}} \sum_{k \in I_{\hat{r}}} \mathcal{M}\left(\frac{\vartheta\left(y_k' - y_G', \frac{m}{2} \right)}{\rho} \right) < \varepsilon\} \in F(I)$$

Conversely, choose $G \in B$. Then

$$\frac{1}{h_{\hat{r}}} \sum_{k \in I_{\hat{r}}} \mathcal{M}\left(\frac{\eta\left(y_k' - l, \frac{m}{2} \right)}{\rho} \right) \leq 1 - \varepsilon \text{ or } \frac{1}{h_{\hat{r}}} \sum_{k \in I_{\hat{r}}} \mathcal{M}\left(\frac{\vartheta\left(y_k' - l, \frac{m}{2} \right)}{\rho} \right) \geq \varepsilon.$$

Now, we show that there exists a number $G = G(y, \varepsilon, m)$ such that

$$\left\{ \hat{r} \in \mathbb{N} : \frac{1}{h_{\hat{r}}} \sum_{k \in I_{\hat{r}}} \mathcal{M}\left(\frac{\eta(y_k' - y_N', m)}{\rho} \right) \leq 1 - n \text{ or} \frac{1}{h_{\hat{r}}} \sum_{k \in I_{\hat{r}}} \mathcal{M}\left(\frac{\vartheta(y_k' - y_N', m)}{\rho} \right) \geq n \right\} \in I.$$

Define for each $y \in \tilde{L}^{I,\theta}_{(\eta,\vartheta)}(\mathcal{M})$,

$$D = \left\{ \hat{r} \in \mathbb{N} : \frac{1}{h_{\hat{r}}} \sum_{k \in I_{\hat{r}}} \mathcal{M}\left(\frac{\eta(y_k' - y_N', m)}{\rho} \right) \leq 1 - n \text{ or } \frac{1}{h_{\hat{r}}} \sum_{k \in I_{\hat{r}}} \mathcal{M}\left(\frac{\vartheta(y_k' - y_N', m)}{\rho} \right) \geq n \right\} \in I.$$

Here we show that $D \subset B$. Let, if possible, $D \not\subseteq B$. Then there exists $\upsilon \in D$ and $\upsilon \notin B$. Thus, we have

$$\frac{1}{h_{\hat{r}}} \sum_{k \in I_{\hat{r}}} \mathcal{M}\left(\frac{\eta(y_k' - y_G', m)}{\rho} \right) \leq 1 - n$$

$$\text{and} \frac{1}{h_{\hat{r}}} \sum_{k \in I_{\hat{r}}} \mathcal{M}\left(\frac{\eta\left(y_k' - l, \frac{m}{2} \right)}{\rho} \right) > 1 - \varepsilon$$

In particular, $\frac{1}{h_{\hat{r}}}\sum_{k\in I_{\hat{r}}}\mathcal{M}\left(\frac{\eta\left(y'_G - l, \frac{m}{2}\right)}{\rho}\right) > 1-\varepsilon$. Thus, we have

$$1-n \geq \frac{1}{h_{\hat{r}}}\sum_{k\in I_{\hat{r}}}\mathcal{M}\left(\frac{\eta\left(y'_\upsilon - y'_G, m\right)}{\rho}\right)$$

$$\geq \frac{1}{h_{\hat{r}}}\sum_{k\in I_{\hat{r}}}\mathcal{M}\left(\frac{\eta\left(y'_\upsilon - l, \frac{m}{2}\right) * \eta\left(y'_G - l, \frac{m}{2}\right)}{\rho}\right)$$

$$\geq (1-\varepsilon)(1-\varepsilon) > 1-n,$$

which is not possible. On the other hand,

$$\frac{1}{h_{\hat{r}}}\sum_{k\in I_{\hat{r}}}\mathcal{M}\left(\frac{\vartheta\left(y'_k - y'_G, m\right)}{\rho}\right) \geq n$$

$$\text{and} \frac{1}{h_{\hat{r}}}\sum_{k\in I_{\hat{r}}}\mathcal{M}\left(\frac{\vartheta\left(y'_k - l, \frac{m}{2}\right)}{\rho}\right) < \varepsilon$$

In particular $\frac{1}{h_{\hat{r}}}\sum_{k\in I_{\hat{r}}}\mathcal{M}\left(\frac{\vartheta\left(y'_G - l, \frac{m}{2}\right)}{\rho}\right) < \varepsilon$. Therefore, we have

$$n \leq \frac{1}{h_{\hat{r}}}\sum_{k\in I_{\hat{r}}}\mathcal{M}\left(\frac{\vartheta\left(y'_\upsilon - y'_G, m\right)}{\rho}\right)$$

$$\leq \frac{1}{h_{\hat{r}}}\sum_{k\in I_{\hat{r}}}\mathcal{M}\left(\frac{\vartheta\left(y'_\upsilon - l, \frac{m}{2}\right) \Diamond \vartheta\left(y'_G - l, \frac{m}{2}\right)}{\rho}\right)$$

$$\leq \varepsilon \Diamond \varepsilon < n,$$

which is not possible. Thus, $D \subset B$.

10.3 INTUITIONISTIC GENERALIZED LUCAS STATISTICAL CONVERGENT SEQUENCE SPACES

In this section, we introduce generalized Lucas statistical completeness with respect to an intuitionistic fuzzy normed space.

Definition 10.14

A sequence $y = (y_k)$ is said to be generalized Lucas statistically convergent (or $\tilde{L}-$ statistically convergence) to l, if for every $\varepsilon > 0$, the set $K_\varepsilon(\tilde{L}) = \{k \le n : |y_k^{'} - l| \ge \varepsilon\}$ has natural density zero, i.e., $d(K_\varepsilon(\tilde{L})) = 0$, i.e.,

$$\lim_{n\to\infty} \frac{1}{n}\left|\left\{k \le n : |y_k^{'} - l| \ge \varepsilon\right\}\right| = 0,$$

where $y_k^{'} = \tilde{L}y_k$ and it is denoted by $d(\tilde{L}) - \lim y_k = l$.

Definition 10.15

A sequence $y = (y_k)$ is said to be $\tilde{L}-$ statistically Cauchy if there exists a number $G = G(\varepsilon)$ such that

$$\lim_{n\to\infty} \frac{1}{n}\left|\left\{k \le n : |y_k^{'} - y_G^{'}| \ge \varepsilon\right\}\right| = 0$$

for every $\varepsilon > 0$.

Definition 10.16

Let $(U, \eta, \vartheta, *, \Diamond)$ be an IFNS. A sequence $y = (y_k)$ is said to be generalized Lucas statistically convergent with respect to IFN (η, ϑ) (or $\tilde{L}$ SC-IFN) to l, if for every $\varepsilon > 0, m > 0$, the set $\{k \le n : \eta(y_k^{'} - l, m) \le 1 - \varepsilon \ or \ \vartheta(y_k^{'} - l, m) \ge \varepsilon\}$ has natural density zero. That is

$$\lim_{n} \frac{1}{n}\left|\left\{k \le n : \eta(y_k^{'} - l, m) \le 1 - \varepsilon \ or \ \vartheta(y_k^{'} - l, m) \ge \varepsilon\right\}\right| = 0.$$

It is denoted by $d(\tilde{L})_{\text{IFN}} - \lim y_k = l$ or $y_k \to l(S(\tilde{L})_{\text{IFN}})$.

Definition 10.17

Let $(U,\eta,\vartheta,*,\Diamond)$ be an IFNS. A sequence $y=(y_k)$ is said to be generalized Lucas statistically Cauchy with respect to IFN (η,ϑ) (or $\tilde{L}$ SCa-IFN), if for every $\varepsilon>0, m>0$, there exists $G=G(\varepsilon)$ such that

$$\left\{k\le n:\eta\left(y_k^{'}-y_G^{'},m\right)\le 1-\varepsilon \ or\ \vartheta\left(y_k^{'}-y_G^{'},m\right)\ge\varepsilon\right\}$$

has natural density zero. That is

$$\lim_n\frac{1}{n}\left|\left\{k\le n:\eta\left(y_k^{'}-y_G^{'},m\right)\le 1-\varepsilon \ or\ \vartheta\left(y_k^{'}-y_G^{'},m\right)\ge\varepsilon\right\}\right|=0.$$

Lemma 10.1

Let $(U,\eta,\vartheta,*,\Diamond)$ be an IFNS. Thus, for every $\varepsilon>0, m>0$, the following statements are equivalent:

i. $d(\tilde{L})_{IFN}-\lim y_k=l.$

ii. $\lim_n\frac{1}{n}\left|\left\{k\le n:\eta\left(y_k^{'}-l,m\right)\le 1-\varepsilon\right\}\right|=\lim_n\frac{1}{n}\left|\left\{\vartheta\left(y_k^{'}-l,m\right)\ge\varepsilon\right\}\right|=0.$

iii. $\lim_n\frac{1}{n}\left|\left\{k\le n:\eta\left(y_k^{'}-l,m\right)>1-\varepsilon\right\}\right|$ and $\lim_n\frac{1}{n}\left|\left\{\vartheta\left(y_k^{'}-l,m\right)<\varepsilon\right\}\right|=1.$

iv. $\lim_n\frac{1}{n}\left|\left\{k\le n:\eta\left(y_k^{'}-l,m\right)>1-\varepsilon\right\}\right|=\lim_n\frac{1}{n}\left|\left\{\vartheta\left(y_k^{'}-l,m\right)<\varepsilon\right\}\right|=1.$

v. $S-\lim\eta\left(y_k^{'}-l,m\right)=1$ and $S-\lim\vartheta\left(y_k^{'}-l,m\right)=0.$

Theorem 10.6

Let $(U,\eta,\vartheta,*,\Diamond)$ be an IFNS. Then, $d(\tilde{L})_{\text{IFN}}-\lim y_k=l$ is unique, if the sequence $y=(y_k)$ is $\tilde{L}$ SC-IFN.

Proof: Suppose $l_1\ne l_2$. Consider that $d(\tilde{L})_{\text{IFN}}-\lim y_k=l_1$ and $d(\tilde{L})_{\text{IFN}}-\lim y_k=l_2$. For any $\varepsilon>0$, take $u>0$ such that $(1-u)*(1-u)>1-\varepsilon$ and $u\Diamond u<\varepsilon$. Define the following sets, for any $m>0$, we have

$$K_{\eta,1}(u,m)=\left\{k\in\mathbb{N}:\eta\left(y_k^{'}-l_1,\frac{m}{2}\right)\le 1-u\right\},$$

$$K_{\eta,2}(u,m)=\left\{k\in\mathbb{N}:\eta\left(y_k^{'}-l_2,\frac{m}{2}\right)\le 1-u\right\},$$

$$K_{\vartheta,1}(u,m)=\left\{k\in\mathbb{N}:\vartheta\left(y_k'-l_1,\frac{m}{2}\right)\geq u\right\},$$

$$K_{\vartheta,1}(u,m)=\left\{k\in\mathbb{N}:\vartheta\left(y_k'-l_2,\frac{m}{2}\right)\geq u\right\}.$$

Thus, for all $m>0$ and using Lemma 10.1, we have

$$K_{\eta,1}(\varepsilon,m)=K_{\vartheta,1}(\varepsilon,m)=0$$

because $d(\tilde{L})_{\text{IFN}}-\lim y_k=l_1$. Similarly, for all $m>0$ we have

$$K_{\eta,2}(\varepsilon,m)=K_{\vartheta,2}(\varepsilon,m)=0$$

because $d(\tilde{L})_{\text{IFN}}-\lim y_k=l_2$. Now, consider

$$K_{\eta,\vartheta}(\varepsilon,m)=\left\{K_{\eta,1}(\varepsilon,m)\cup K_{\eta,2}(\varepsilon,m)\right\}\cap\left\{K_{\vartheta,1}(\varepsilon,m)\cup K_{\vartheta,2}(\varepsilon,m)\right\}.$$

Therefore, we observe that $d\left(K_{\eta,\vartheta}(\varepsilon,m)\right)=0$ which implies that $d\left(\mathbb{N}\backslash K_{\eta,\vartheta}(\varepsilon,m)\right)=1$. If $k\in\mathbb{N}\backslash K_{\eta,\vartheta}(\varepsilon,m)$, then we have two possible cases:

i. $k\in\mathbb{N}\backslash\left(K_{\eta,1}(\varepsilon,m)\cup K_{\eta,2}(\varepsilon,m)\right)$,

ii. $k\in\mathbb{N}\backslash\left(K_{\vartheta,1}(\varepsilon,m)\cup K_{\vartheta,2}(\varepsilon,m)\right)$.

First we assume that the condition (i) holds. Then, we have

$$\eta(l_1-l_2,m)\geq\eta\left(y_k'-l_1,\frac{m}{2}\right)*\eta\left(y_k'-l_2,\frac{m}{2}\right)>(1-u)*(1-u).$$

In this case,

$$\eta(l_1-l_2,m)>1-\varepsilon, \tag{10.1}$$

because $(1-u)*(1-u)>1-\varepsilon$. By using (10.1), for all $m>0$, we get $\eta(l_1-l_2,m)=1$, where $\varepsilon>0$ is arbitrary, which implies $l_1=l_2$.

Moreover, if $k\in\mathbb{N}\backslash\left(K_{\vartheta,1}(\varepsilon,m)\cup K_{\vartheta,2}(\varepsilon,m)\right)$, then we write

$$\vartheta(l_1-l_2,m)\leq\vartheta\left(y_k'-l_1,\frac{m}{2}\right)\Diamond\vartheta\left(y_k'-l_2,\frac{m}{2}\right)<u\Diamond u.$$

Using $u \lozenge u < \varepsilon$, we have

$$\vartheta(l_1 - l_2, m) < \varepsilon.$$

For all $m > 0$, we get $\vartheta(l_1 - l_2, m) = 0$, where $\varepsilon > 0$ is arbitrary, which implies $l_1 = l_2$.

Theorem 10.7

Let $(U, \eta, \vartheta, *, \lozenge)$ be an IFNS. Then $d(\tilde{L})_{\text{IFN}} - \lim y_k = l$ if there exists a subset $J = \{j_1, j_2, \cdots\} \subset \mathbb{N}$, such that $d(J) = 1$ and $(\eta, \vartheta) - \lim_{n \to \infty} y_{j_n} = l$.

Proof: Suppose that $d(\tilde{L})_{\text{IFN}} - \lim y_k = l$. For any $m > 0$ and $v = 1, 2, \cdots$

$$M_{\eta,\vartheta}(v,m) = \left\{ k \in \mathbb{N} : \eta\left(y_k' - l, m\right) > 1 - \frac{1}{v} \text{ or } \vartheta\left(y_k' - l, m\right) < \frac{1}{v} \right\}$$

and

$$P_{\eta,\vartheta}(v,m) = \left\{ k \in \mathbb{N} : \eta\left(y_k' - l, m\right) \le 1 - \frac{1}{v} \text{ or } \vartheta\left(y_k' - l, m\right) \ge \frac{1}{v} \right\}.$$

Then, $d\left(P_{\eta,\vartheta}(v,m)\right) = 0$, because $d(\tilde{L})_{\text{IFN}} - \lim y_k = l$. Further, for $t > 0$ and $v = 1, 2, \ldots,$

$$M_{\eta,\vartheta}(v,m) \supset M_{\eta,\vartheta}(v+1,m)$$

and

$$d\left(M_{\eta,\vartheta}(v,m)\right) = 1 \tag{10.2}$$

for $m > 0$ and $v = 1, 2, \cdots$.

Now, we will show that for $k \in M_{\eta,\vartheta}(v,m),\ y_k \to l(S\left(\tilde{L}\right)_{\text{IFN}})$. Consider for some $k \in M_{\eta,\vartheta}(v,m),\ y_k \nrightarrow l(S\left(\tilde{L}\right)_{\text{IFN}})$. Thus there exists $\beta > 0$ and a positive integer k_0 such that for all $k \ge k_0$,

$$\eta\left(y_k' - l, m\right) \le 1 - \beta \text{ or } \vartheta\left(y_k' - l, m\right) \ge \beta.$$

Let $\eta\left(y_k' - l, m\right) > 1 - \beta$ or $\vartheta\left(y_k' - l, m\right) < \beta$, for all $k \ge k_0$. Then

$$d\left(\left\{k \in \mathbb{N} : \eta\left(y_k' - l, m\right) > 1 - \beta \text{ or } \vartheta\left(y_k' - l, m\right) < \beta\right\}\right) = 0.$$

Since $v > \frac{1}{v}$, we have $d\left(M_{\eta,\vartheta}(v,m)\right) = 0$ which contradicts (10.2). Thus, $k \in M_{\eta,\vartheta}(v,m),\ y_k \to l(S\left(\tilde{L}\right)_{\text{IFN}})$.

Consider there exists a subset $J = \{j_1, j_2, \ldots\} \subseteq \mathbb{N}$, such that $d(J) = 1$ and $(\eta, \vartheta) - \lim_{n \to \infty} y_{j_n} = l$, i.e., there exists $G \in \mathbb{N}$ such that for every $\beta > 0$ and $m > 0$,

$$\eta\left(y_k' - l, m\right) > 1 - \beta \text{ or } \vartheta\left(y_k' - l, m\right) < \beta.$$

Now,

$$K_{\eta,\vartheta}(\beta, m) = \left\{k \in \mathbb{N} : \eta\left(y_k' - l, m\right) \le 1 - \beta \text{or} \vartheta\left(y_k' - l, m\right) \ge \beta\right\} \subseteq \mathbb{N} - \left\{j_{G+1}, j_{G+2}, \ldots\right\}.$$

Thus $d\left(M_{\eta,\vartheta}(\beta, m)\right) \le 1 - 1 = 0$. Hence $d(\tilde{L})_{\text{IFN}} - \lim y_k = l$.

Theorem 10.8

Let $(U, \eta, \vartheta, *, \Diamond)$ be an IFNS. If a sequence $y = (y_k)$ is $\tilde{L}\text{SC} - \text{IFN}$, then it is $\tilde{L}\text{SCa} - \text{IFN}$.

Proof: Suppose $d(\tilde{L})_{\text{IFN}} - \lim y_k = l$. For given $\varepsilon > 0$, choose $v > 0$ such that $(1-\varepsilon)*(1-\varepsilon) > 1 - v$ and $\varepsilon \Diamond \varepsilon < v$. Then for $m > 0$, we have

$$d\left(A(\varepsilon, m)\right) = d\left(\left\{k \in \mathbb{N} : \eta\left(y_k' - l, \frac{m}{2}\right) \le 1 - \varepsilon \text{ or } \vartheta\left(y_k' - l, \frac{m}{2}\right) \ge \varepsilon\right\}\right) = 0 \quad (10.3)$$

which implies that

$$d\left(A^c(\varepsilon, m)\right) = d(\{k \in \mathbb{N} : \eta\left(y_k' - l, \frac{m}{2}\right) > 1 - \varepsilon \text{ or } \vartheta\left(y_k' - l, \frac{m}{2}\right) < \varepsilon\}) = 1.$$

Let $p \in A^c(\varepsilon, m)$. Then

$$\eta\left(y_p' - l, m\right) > 1 - \varepsilon \text{ or } \vartheta\left(y_p' - l, m\right) < \varepsilon.$$

Now, suppose

$$B(\varepsilon, m) = \left\{k \in \mathbb{N} : \eta\left(y_k' - y_p', m\right) \le 1 - v \text{ or } \vartheta\left(y_k' - y_p', m\right) \ge v\right\}.$$

We need to show that $B(\varepsilon, m) \subset A(\varepsilon, m)$. Let $k \in B(\varepsilon, m) / A(\varepsilon, m)$. Thus, we have

$$\eta\left(y_k' - y_p', m\right) \le 1 - v \text{ and } \eta\left(y_k' - l, \frac{m}{2}\right) > 1 - \varepsilon,$$

in particular $\eta\left(y_q' - l, \frac{m}{2}\right) > 1 - \varepsilon$. Then

$$1 - v \ge \eta\left(y_k' - y_p', m\right) \ge \eta\left(y_k' - l, \frac{m}{2}\right) * \eta\left(y_q' - l, \frac{m}{2}\right) > (1-\varepsilon)*(1-\varepsilon) > 1 - v,$$

which is not possible. On the other hand,

$$\vartheta\left(y_k' - y_p', m\right) \ge v \text{ and } \vartheta\left(y_k' - l, \frac{m}{2}\right) < \varepsilon,$$

in particular $\vartheta\left(y_q^{'}-l,\frac{m}{2}\right)<\varepsilon$. Then

$$v\le\vartheta\left(y_k^{'}-y_p^{'},m\right)\le\vartheta\left(y_k^{'}-l,\frac{m}{2}\right)\Diamond\vartheta\left(y_q^{'}-l,\frac{m}{2}\right)<\varepsilon\Diamond\varepsilon<v,$$

which is not possible. Thus, $B(\varepsilon,m)\subset A(\varepsilon,m)$. Therefore by (10.3) $d\left(B(\varepsilon,m)\right)=0$. Hence, y is $\tilde{L}$ SCa-IFN.

Definition 10.18. An IFNS $(U,\eta,\vartheta,*,\Diamond)$ is said to be complete (Saadati and Park 2006) if every Cauchy sequence is convergent in $(U,\eta,\vartheta,*,\Diamond)$.

Definition 10.19

An IFNS $(U,\eta,\vartheta,*,\Diamond)$ is said to be statistically ($\tilde{L}$ SC-IFN) complete if every statistically ($\tilde{L}$ SC-IFN, respectively) Cauchy sequence with respect to intuitionistic fuzzy norm (η,ϑ) is statistically ($\tilde{L}$ SC-IFN, respectively) convergent with respect to intuitionistic fuzzy norm (η,ϑ).

Theorem 10.9

An IFNS$(U,\eta,\vartheta,*,\Diamond)$ is $\left(\tilde{L}\text{SC}-\text{IFN}\right)$– complete.

Proof: Suppose $y=(y_k)$ be ($\tilde{L}$ SC-IFN)-Cauchy but not ($\tilde{L}$ SC-IFN)-convergent. For given $\varepsilon>0$ and $m>0$, choose $v>0$ such that $(1-\varepsilon)*(1-\varepsilon)>1-v$ and $\varepsilon\Diamond\varepsilon<v$. Now,

$$\eta\left(y_k^{'}-y_G^{'},m\right)\ge\eta\left(y_k^{'}-l,\frac{m}{2}\right)*\eta\left(y_G^{'}-l,\frac{m}{2}\right)>(1-\varepsilon)*(1-\varepsilon)>1-v$$

and

$$\vartheta\left(y_k^{'}-y_G^{'},m\right)\le\vartheta\left(y_k^{'}-l,\frac{m}{2}\right)\Diamond\vartheta\left(y_G^{'}-l,\frac{m}{2}\right)<\varepsilon\Diamond\varepsilon<v,$$

since y is not $\left(\tilde{L}\text{SC}-\text{IFN}\right)$– convergent. Therefore, $d\left(F^c(\varepsilon,m)\right)=0$, where

$$F(\varepsilon,m)=\left\{k\in\mathbb{N}:\vartheta_{y_k^{'}-y_G^{'}}(\varepsilon)\le1-s\right\}$$

and so $d\left(F(\varepsilon,m)\right)=1$, which is a contradiction, since y is ($\tilde{L}$ SC-IFN)-Cauchy, so that y must be ($\tilde{L}$ SC-IFN)-convergent. Hence, every IFNS is ($\tilde{L}$ SC-IFN)-complete.

Now, using Theorems 10.7–10.9, we immediately deduce the following result:

Theorem 10.10

Let $(U,\eta,\vartheta,*,\Diamond)$ be an *IFNS*. Then, for any sequence $y=(y_k)\in U$, the following conditions are equivalent:

i. y is ($\tilde{L}$ SC-IFN)-convergent.
ii. y is ($\tilde{L}$ SC-IFN)-Cauchy.
iii. IFNS$(U,\eta,\vartheta,*,\Diamond)$ is ($\tilde{L}$ SC-IFN)-complete.
iv. There exists an increasing index sequence $K=(k_n)$ of natural numbers such that $d(K)=1$ and the subsequence (y_{k_n}) is a ($\tilde{L}$ SC-IFN)-Cauchy.

10.4 CONCLUSION

Several mathematicians studied various properties related to generalized Lucas matrix. But we aimed to introduce new sequence spaces $\tilde{L}^{I,\theta}_{(\eta,\vartheta)}(\mathcal{M}),\tilde{L}^{I,\theta}_{0(\eta,\vartheta)}(\mathcal{M})$ and studied some algebraic and topological properties. Also we studied generalized Lucas statistical convergence ($\tilde{L}$ SC-IFNS) and generalized Lucas statistical Cauchy sequences ($\tilde{L}$ SCa-IFNS) with respect to an IFNS. ($\tilde{L}$ SC-IFNS) completeness property is also given in this chapter. As future work, one can study these results by using modulus function, generalized Fibonacci difference spaces, etc.

REFERENCES

Atanassov, K. T. 1986. Intuitionistic fuzzy sets. *Fuzzy Sets Syst.* 20: 87–96.

Belen, C., and S. A. Mohiuddine. 2013. Generalized weighted statistical convergence and application. *Appl. Math. Comput.* 219: 9821–9826.

Braha, N. L., H. M. Srivastava, and S. A. Mohiuddine. 2014. A Korovkin's type approximation theorem for periodic functions via the statistical summability of the generalized de la Vallée Poussin mean. *Appl. Math. Comput.* 228: 162–169.

Connor, J. S. 1988. The statistical and strong *p*-Cesàroconvergence of sequence. *Analysis* 8: 47–63.

Demiriz, S., and H. B. Ellidokuzoglu. 2018. On some new paranormed Lucas sequences paces and Lucas core. *Conf. Proc. Sci. Technol.* 1(1): 32–35.

Dems, K. 2004. On *I*-Cauchy sequences. *Real Anal. Exchange* 30: 123–128.

Fast, H. 1951. Sur la convergence statistique. *Colloq. Math.* 2: 241–244.

Freedman, A. R., J. J. Sember, and M. Raphael. 1978. Some Cesàro-type summability spaces. *Proc. London Math. Soc.* 37: 508–520.

Fridy, J. A. 1985. On statistical convergence. *Analysis* 5(4): 301–313.

Fridy, J. A., and C. Orhan. 1993. Lacunary statistical convergence. *Pac. J. Math.* 160: 43–51.

Hazarika, B., A. Alotaibi, and S. A. Mohiuddine. 2020. Statistical convergence in measure for double sequences of fuzzy-valued functions. *Soft Computing* 24: 6613–6622.

Kadak, U., and S. A. Mohiuddine. 2018. Generalized statistically almost convergence based on the difference operator which includes the (p, q)-gamma function and related approximation theorems. *Results Math.* 73(1): Article 9.

Kadak, U., M. Mursaleen, and S. A. Mohiuddine. 2019. Statistical weighted matrix summability of fuzzy mappings and associated approximation results. *J. Intell. Fuzzy Syst.* 36: 3483–3494.

Karakaya, V., N. Şimşek, M. Ertürk, and F. Gürsoy. 2012. Statistical convergence of sequences of functions in intuitionistic fuzzy normed spaces. *Abstr. Appl. Anal.* Volume 2012, Article ID 157467, 19 pages.

Karakuş, S., K. Demirci, and O. Duman. 2008. Statistical convergence on intuitionistic fuzzy normed spaces. *Chaos Solutions Fractal* 35: 763–769.

Karakas, M., and A. M. Karakas. 2018. A study on Lucas difference sequence spaces and ℓ_p $(E(r,s))$ and $\ell_\infty(E(r,s))(r,s))\}$. *Maejo Int. J. Sci. Technol.* 2(1): 70–78.

Khan, V. A., F. H. Yasmeen, H. Altaf and Q. D. Lohani. 2016. Intuitionistic fuzzy I-convergent sequence spaces defined by compact operator. *Cogent Math.* 3: 1267904.

Koshy, T. 2001. *Fibonacci and Lucas Numbers with Applications.* John Wiley & Sons, Hoboken, NJ.

Kostyrko, P., T. Salat, and W. Wilczynski. 2000. I-convergence. *Real Anal. Exchange* 26(2):669–686.

Lael, F., and K. Nourouzi. 2008. Some results on the IF-normed spaces. *Chaos Solitons and Fractals* 37: 931–939.

Lindenstrauss, J., and L. Tzafriri. 1971. An Orlicz sequence spaces. *Israel J. Math.* 10: 379–390.

Mohiuddine, S. A., and B. A. S. Alamri. 2019. Generalization of equi-statistical convergence via weighted lacunary sequence with associated Korovkin and Voronovskaya type approximation theorems. *Rev. R. Acad. Cienc. Exactas Fís. Nat. Ser. A Math. RACSAM* 113(3): 1955–1973.

Mohiuddine, S. A., A. Asiri, and B. Hazarika. 2019. Weighted statistical convergence through difference operator of sequences of fuzzy numbers with application to fuzzy approximation theorems. *Int. J. Gen. Syst.* 48(5): 492–506.

Mohiuddine, S.A., and B. Hazarika. 2017. Some classes of ideal convergent sequences and generalized difference matrixoperator. *Filomat* 31(6): 1827–1834.

Mohiuddine, S. A., B. Hazarika, and M. A. Alghamdi. 2019. Ideal relatively uniform convergence with Korovkin and Voronovskaya types approximation theorems. *Filomat* 33(14): 4549–4560.

Mohiuddine, S. A., and Q. M. D. Lohani. 2009. On generalized statistical convergence in intuitionistic fuzzy normed space. *Chaos Solitons Fractals* 42: 1731–1737.

Mohiuddine, S. A., K. Raj, M. Mursaleen, and A. Alotaibi. 2020. Linear isomorphic spaces of fractional-order difference operators. *Alexandria Eng. J* Doi: 10.1016/j.aej.2020.10.039.

Mursaleen, M., V. Karakaya, and S. A. Mohiuddine. 2010. Schauder basis, separability, and approximation property in intuitionistic fuzzy normed space. *Abstr. Appl. Anal.* 2010: 1–14, Article ID 131868.

Mursaleen, M., and Q. M. D. Lohani. 2009. Intuitionistic fuzzy 2-normed space and some related concepts. *Chaos Solitons Fractals* 42: 224–234.

Mursaleen, M., and S. A. Mohiuddine. 2009a. Onlacunary statistical convergence with respect to the intuitionistic fuzzy normed space. *J. Comput. Appl. Math.* 233: 142–149.

Mursaleen, M., and S. A. Mohiuddine. 2009b. Statistical convergence of double sequences in intuitionistic fuzzy normed spaces. *Chaos Solitons Fractals* 41: 2414–2421.

Park, J. H. 2004. Intuitionistic fuzzy metric spaces. *Chaos Solitions Fractals* 22: 1039–1046.

Raj, K., A. Choudhary, and C. Sharma. 2018. Almost strongly Orlicz double sequence spaces of regular matrices and their applications to statistical convergence. *Asian-Eur. J. Math.* 11: Article 1850073.

Raj, K., S. Pandoh, and S. Jamwal. 2015. Fibonacci difference sequence spaces for modulus functions. *Matematiche (Catania)* 70: 137–156.

Raj, K., C. Sharma, and A. Choudhary. 2018. Applications of Tauberian theorem in Orlicz spaces of double difference sequences of fuzzy numbers. *J. Intell. Fuzzy Syst.* 35: 2513–2524.

Saadati, R., and J. H. Park. 2006. On the intuitionistic fuzzy topological spaces. *Chaos Solitions Fractals* 27: 331–344.

Salat, T. 1980. On statistical Convergent sequence of fuzzy real numbers. *Math. Slovaca.* 45: 269–273.

Schweizer, B., and A. Sklar. 1960. Statistical metric spaces. *Pacific J. Math.* 10: 314–344.

Steven, V. 2008. *Fibonacci and Lucas Numbers, and the Golden Section Theory and Applications.* Dover, US.

Tripathy, B. C., and B. Hazarika. 2011. Some I-convergent sequence spaces defined by Orlicz functions. *Acta Math. Appl. Sin. Engl. Ser.* 27(1): 149–154.

Tripathy, B. C., and B. Hazarika and B. Choudhary. 2012. Lacunary I-convergent sequences. *Kyungpook Math. J.* 52(4): 473–482.

Yaying, T., and B. Hazarika. 2019. Lacunary arithmetic convergence. *Proc. Jangjeon Math. Soc.* 21 (3): 507–513.

Yaying, T., and B. Hazarika. 2020. Lacunary arithmetic statistical convergence. *Natl. Acad. Sci. Lett.* 43: 547–551.

Yaying, T., B. Hazarika, S.A. Mohiuddine, M. Mursaleen, and K. J. Ansari. 2020. Sequence spaces derived by the triple band generalized Fibonacci difference operator. *Adv. Difference Equ.* 2020: Article 639.

Zadeh, L. A. 1965. Fuzzy set. *Inf. Control* 8: 338–353.

11 Soft Computing Techniques in Social Sciences: The Recent Developments

Anuradha Bhattacharjee
Assam University

CONTENTS

11.1 INTRODUCTION

Change is universal, and with change comes gradual development. The world has witnessed a rapid development in the prospect of utilisation of different aspects because of technological advancements. With the advent of technological innovations, soft computing techniques, also, emerged as a non-conventional approach towards finding solution for problems inappropriate for conventional technical ways of solving it.

The paradigm of soft computing encompasses mathematical techniques in solving the complex real-life problems which are capable of handling uncertainty and partial truth. As a matter of fact, soft computing techniques do not adhere to strict mathematical and technical dimensions, and as such, human mind is given importance to.

Because of its adaptive nature, soft computing technique is predominant in social sciences, too, as it renders human knowledge in the form of 'cognition, recognition and learning'. Since soft computing techniques are indifferent towards strict mathematical formulation, social science arena of recent times focus on the usage of it, as the characteristics of the human brain to recognise and perceive can be deduced through it. Behavioural approach can be studied through probabilistic reasoning, and diverse solutions can be provided for one complex problem. The present study is an attempt towards analysing the emergence of soft computing techniques in the field of social sciences, which has the ability to structure hazy problems which are fuzzy in nature amidst the environment of uncertainty.

11.2 ON CONTEMPLATING ABOUT PROBLEM SOLVING

The world is a hub of technical problems. There are certain problems which cannot be solved logically but only by theorising, which, however, is difficult because a big resource and much time are needed for computing, and this can be done effectively through the methods of nature, which in turn results in an optimal solution if not a mathematical solution (Das et al. 2013).

The word computer can be said to have come from 'compute' meaning 'calculation'. COMPUTER, however, stands for 'Common Operating Machine Particularly Used for Technical Education and Research' (Youth4work 2020). To begin with, the whole process can be understood with the meaning of the very term 'computing' which is 'the process to transform the input into our required output by control actions' (Youth4work 2020). Das et al. (2013) and Chandana (2019) demonstrated the problem-solving techniques of computing. This computing can be categorised into two sections – hard computing and soft computing. Hard computing consists of precise model which is, again, categorised into symbolic logic reasoning and traditional numerical modeling and search.

11.3 ON DEFINING SOFT COMPUTING

The term 'soft computing' gained prominence in the nineties. Lotfi Zadeh discussed about soft computing and introduced the very terms 'soft computing' and 'fuzzy sets' in his renowned speech on the topic 'A New View on System Theory' in the seminar held in Brooklyn (Seising and Sanz 2020). It is an approach. According to Dogan Ibrahim (2016), soft computing is important as it solves the 'complex real-life problems'. It uses approximate models, and its role model is the mind of man (Ibrahim 2016).

According to McGraw-Hill Dictionary of Scientific and Technical Terms (2003), soft computing is defined as a family of different methods which is just the replica of human brain and which learns to function just like a living person. L. A. Zadeh (1996), the man behind the emergence of soft computing defined it as a cluster of methodologies whose 'role model' is the human mind. In fact, soft computing is 'the science of reasoning' which just like an umbrella undertakes different computational techniques to solve the constituent problem (Das et al. 2013).

The 'cognitive behaviour' of our mind can be structured through soft computing, and soft computing itself is the structure of 'conceptual intelligence' of the machineries (Chandana 2019). When hard computing fails to give the desired output, soft computing is the last resort. Seising and Sanz (2012) maintained that hard computing is applied to get the appropriate and exact result of any particular problem, but there are certain problems which cannot get solution in such a way, and this is when soft computing becomes effective.

11.4 ON DESCRIBING FEATURES OF SOFT COMPUTING

The characteristics of soft computing can be better understood by the following Table 11.1:

11.5 ON DIFFERENTIATING BETWEEN HARD COMPUTING AND SOFT COMPUTING

The difference between hard computing and soft computing is quite impressive.

Rudolf Seising and Veronica Sanz (2012) analysed the differences that exist between hard computing and soft computing.

According to them,

> Hard computing is rigid and is precisely done, whereas soft computing is flexible and has approximation.
>
> If hard computing is bi-valued, soft computing depends on fuzzy value.
>
> There is total order in hard computing. On the other hand, there is partial order in soft computing.
>
> If hard computing is based on abstract ideas, soft computing is empirically structured.
>
> Hard computing is unique, but soft computing justifies hybridisation.
>
> Hard computing points towards numbers, whilst soft computing points towards words.

TABLE 11.1
Features of Soft Computing

1	Solve complex day-to-day problems
2	Resistant towards partial truth
3	Tolerate imprecision
4	Approximation
5	Cost-effective
6	Miniature of human intelligence
7	Miniature of computational intelligence
8	Adaptive
9	Consists of both natural and artificial ideas (Das et al. 2013)
10	'Extension of natural heuristics' (Das et al. 2013)

11.6 ON ANALYSING SOFT COMPUTING TECHNIQUES

Das et al. (2013) described in detail the division of soft computing techniques. It is categorised into approximate reasoning and functional approximation and randomised search. Approximate reasoning is, again, classified into probabilistic model and fuzzy logic, and the other one is classified into evolutionary computing and neural network. Evolutionary computing is divided into – evolution strategy, evolutionary programming, genetic algorithm and genetic programming. Neural network, again, has feed forward neural network and recurrent neural network. Probabilistic model has Bayesian and Shafer theory, i.e., probability of fuzzy event and belief of fuzzy event.

11.6.1 Artificial Neural Network

Neural network is commendable in solving the problems related to classification, but it is not helpful in critical applications. It has gained prominence in 'pattern recognition'. It has, also, emerged as a tool of classification of 'remotely sensing images'. Neural network has also received attention for generalising from unknown truths, and it also makes inferences from present examples. It requires no statistical computation of mean and standard deviation (Jindal and Jindal 2012).

Artificial intelligence is constructed by artificial neural network, and this artificial neural network is constructed just like the way the human mind analyses a problem. Problems which cannot be solved by a person or through statistical measures can be analysed by it (Frankenfield 2020).

Neurons constitute artificial neural networks, and the neurons pass signals which travel from the input layer to the output later systematically (Wikipedia 2020a).

11.6.2 Genetic Algorithm

It is an array of metaheuristic which renders strong solutions to optimal problems. It is a subset of evolutionary algorithms, and it should consist of 'genetic representation' and 'fitness function' (Wikipedia 2020a). Genetic algorithm is used for an automatic recognition of neighbourhood, texture, average value of the pixels of spectral channel and known for filtering barriers of feature classification (Jindal and Jindal 2012).

11.6.3 Probabilistic Logic

The amalgamation of probability theory and deductive logic for analysing the uncertain can be referred to as the functioning of the probabilistic logic. Probabilistic logic makes computation much more difficult, although they are derived from logic truth tables (Wikipedia 2020a).

11.6.4 Machine Learning

It is a part of artificial intelligence which is useful in taking past experiences into account to analyse computational algorithms. Computers work on the basis of the data stored instead of being programmed. Machine learning generalises from predictive patterns, whilst in statistics, samples decide the results; both machine learning and statistics are connected with each other in a sense (Wikipedia 2020a).

11.6.5 Bayesian Statistics

It is based on probabilistic reasoning. This is an interpretation of probability based on 'a degree of belief'. The information is represented upon previous knowledge regarding the events which may be either speculation, long run beliefs, observation or experimentation results. However, this method takes much computation into account to find out the answer of statistical problems (Wikipedia 2020a).

11.6.6 Fuzzy Logic

Fuzzy logic is inferential towards the processing of image and 'pattern recognition' as it is functional for the demonstration of uncertainty and partial truths (Jindal and Jindal 2012). Fuzzy logic can be a miniature of human intelligence in performing a particular function. It can render full information for image description as it lessens the inaccuracy of classification. When any information range is neither fully true nor fully false, then fuzzy sets are essential to interpret the vagueness of the knowledge. Fuzzy sets recognise and manipulate events as human minds utilise their decision-making ability on the basis of 'imprecise and non-numerical' data (Wikipedia 2020a).

Zadeh proposed the fuzzy set concept in the year 1965.

11.7 ON DISCUSSING APPLICATIONS OF SOFT COMPUTING TECHNIQUES

The applications of soft computing include:

- In banking area, soft computing helps to determine credit card issues, forgery, bank frauds and online banking frauds.
- Soft computing techniques help in financial market and stock market.
- Soft computing techniques are helpful for face recognition, handwriting identification, image analysis and malwares.
- In the field of medicine, they are applicable for the identification of cancer. Computerised analysis and diagnosis of diseases are possible through soft computing.
- Cyber security is provided through the usage of soft computing techniques. Spam mails are identified and sorted out through the usage of soft computing techniques.
- Soft computing is applicable in the field of aircraft, machineries and artifacts like – washing machines and inverters. They are, also, needed in the operation of metro trains and in the control of railway operation.

These are helpful in proving the theorems and mathematical analysis of fuzzy ideas.
In agriculture, bioinformatics and robotics, too, soft computing techniques have proved to have applications.
In DNA classification and pattern recognition, too, soft computing techniques have proved to have applications.
The process of data mining, also, requires application of soft computer techniques.

Certain research studies have been analysed in the past years on the different application areas of soft computing.

Zadeh (1998) wrote about the applications of soft computing techniques in smart technologies even in that time. He discussed about how software, camera usage and forgeries of online technologies can be analysed through the said techniques.

Gen and Yun (2006) maintained as to how soft computing techniques have emerged as a dynamic discipline in solving the problems incurred in reliability optimisation. The different problems of the same as discussed by the researchers were of 'redundant system', 'alternative design', 'time-dependent' and 'fuzzy'.

Yu and Liu (2006) conducted a study on construction databases. The techniques of symbolic reasoning and numeric reasoning proved to be effective in digging out the data mining. In this paper, the symbolic and numeric techniques in the form of hybridisation provided success in solving the issues of rendering inaccurate knowledge and data and discussed how data mining is helpful in finding out the scarce database.

Yildirim et al. (2011) in their research analysed about earthquake and quarry blast. Their locale of the study was Istanbul and the zone of Marmara. Their aim was to find out the effectiveness of different soft computing techniques to find out the differentiation between earthquake and quarry blast. The model based on peak amplitude ratio and complexity value proved successful on an average of 97.67 percentage to cent percent.

Das et al. (2013) demonstrated the application of soft computing as to how engineering and medical science can combine together in solving the issues of diagnosis and treatment of health ailments. They discussed how soft computing techniques work for different industrial units like steel process, and also how crime can be controlled, and people can be assured of their safety because of the application of soft computing techniques.

Chandana (2019) talked about genetic algorithm. They narrated the function of genetic algorithm in demonstrating the representation of a student's education and the formation of his career – how and what can have an impact on his or her academic performance. They, also, discussed how in music, the music pitches are harmonised by the soft computing techniques and also how fuzzy logic is needed in diagnosing the illness of diabetes and how blood pressure is kept in control in an anesthetic patient.

Falcone et al. (2020) discussed about earthquake engineering as an application area of soft computing technique in their research work.

11.8 ON DISCUSSING ABOUT SOFT COMPUTING TECHNIQUES IN SOCIAL SCIENCE

Seising and Sanz (2012), on discussing about Snow's concept, detailed about the distinction between the 'two cultures' – hard science and soft science. With the passage of time, there emerged a 'third culture' when artificial intelligence merged both of them. Zadeh, on talking about his inclusion of the fuzzy set, expressed his surprise that why then soft computing was not applied in the disciplinary field of law, economics, linguistics and philosophy (Seising and Sanz, 2012). Of late, soft computing techniques have been incorporated in the social science in real day-to-day lives to solve partial truths and uncertainty as fuzzy sets have the same feature of being ambiguous to make soft science and hard science inter-disciplinary in their problem-solving analysis.

As logical principles itself are the basis of fuzziness, Termini (2012) stated how the dilemma between the relative truth and false in logical principles that are unquestioned can be analysed through fuzzy theory. Even the concept of cause-effect is interrelated, and in a study by Puente et al. (2012), different causal and conditional sentences are extracted and analysed through causal graphs. Soft computing has proved its presence in this too.

Soft computing has enumerated importance in the field of science and technology studies and also in the matter of feminist studies of technology. Both of them are involved with artificial intelligence (Sanz 2012). According to her, although critics criticised about the foundation of classical artificial intelligence, however, soft computing has shown intensive changes. Zadeh (2012) talked about the concepts of fuzzy deontic set and rule that provide prescriptions for morality and legality of actions. As fuzzy is qualitative, deontic decision-making is enabled in the fields of ethics and law. When decision-makers cannot make effective decision and the social scientists fail to construct a dynamic approach to problems, fuzzy logic applies qualitative support system. Fuzzy Cognitive Maps and later on Rule Based Fuzzy Cognitive Maps were added to make qualitative dynamic effective in social science (Carvalho 2012).

'Linguistic ambiguity', as discussed by Bradley (2012), can cause communication troublesome as it may incur in misinterpretation of texts said. The patients of the illness of Aphasia cannot rightly interpret the meaning of any language, and the research work by Bradley (2012) analyses the importance of soft computing in reducing the ambiguous nature. The communication process is discussed by Weaver's model where Shannon and Weaver discuss how the sender communicates his message to the receiver of the semantic message (Seising 2012). Gander et al. (2012) in his study discussed about how important fuzzy set theory is when Optical character Recognition is used for digitisation and how information is retrieved about historical texts and how e-books are analysed.

Fuzzy numbers are used to select the required threshold which can help in the process of decision-making regarding whom and how to vote (Lapresta and Piggins 2012).

Some of the applications of soft computing techniques in social science can be listed below:

Different soft computing techniques enable marketing and online advertisements. They are also used in economic modeling.
Linguistic analysis and behavioural analysis can be performed with the implications of some of the soft computing techniques.
Soft computing techniques can be effective in research in social science disciplines.
Certain studies were conducted on the application of soft computing techniques in the field of social sciences.

Ko et al. (2010) found out that soft computing techniques like fuzzy logic and genetic algorithm are useful in the process of supply chain management. They are employed in supply chain management to understand the relationship of an organisation with its customers.

Verikas et al. (2010) analysed the applications of both hybrid and ensemble types of soft computing techniques through a survey, as to how they predict bankruptcy.

Mahdipour and Dadkhah (2012) conducted a study on past history of automatic fire detection from 2000 to 2010. Soft computing techniques are capable of analysing factors that are detrimental towards the natural environment through early detection. The timely detection of fire in uninhabited and inhabited places can be easily predicted using the soft computing techniques.

Chang et al. (2016) in this study found out that soft computing techniques have the capability to predict groundwater levels of a region. The hybrid model of SOM and NARX can regulate sustainable resources. Soft computing techniques make an early prediction of groundwater heights which prove beneficial for the water resource department of a nation.

Kumar and Jaiswal (2017) conducted their study on empirical basis. Soft computing techniques can be used to demonstrate the social platforms Twitter and Tumblr for sentiment analysis, and the researchers conducted estimation on 3000 tweets and 3000 tumblogs of trending issues using soft computing techniques to record the sentiment level. Sentiment problems often arise these days in social media. Support vector machine proved successful to be the most potent in finding sentiment accuracy. Other soft computing techniques are also potent in it.

Bolla and Chakravarthy (2018) conducted a survey to understand the application of social network analysis. Soft computing techniques are usable in understanding the different methods of social network analysis.

Kumar and Sachdeva (2019) conducted a meta-analysis to analyse how soft computing techniques can recognise cyber bullying in the social multi-media. Today, social media has become a powerful platform which is a source of presentation of different activities. Soft computing techniques are applicable in detecting the 'unfavourable' cyber bullying that happens around.

11.9 CONCLUSION

Soft computing as a domain of computational progress encompasses the discipline of pure and applied science to engulf the school of social science and humanities. The soft computing techniques have been positively proved to be potentially powerful to estimate and solve different problems encountered in the field of social science. With due acceleration in the advancement of technology, the soft computing techniques will supposedly emerge as a predominant way as the personification of the human mind towards problem-solving in every kind in the social science arena and, in turn, have a profound impact on the lives of human and the society, as a whole.

REFERENCES

"Artificial Neural Network," Wikipedia, 2020a. last modified November 13, 2020, https://en.wikipedia.org > wiki > Artificial_neural_network.

"Bayesian Statistics," Wikipedia 2020b–2020f. last modified November 14, 2020, https://en.m.wikipedia.org > wiki > Bayesia...

Bolla, J.V., and S.L. Chakravarthy. 2018. Applications of soft computing techniques for social network analysis: A survey. *International Journal of Pure and Applied Mathematics* 120(6): 4237–4258.

Bradley, J. 2012. Syntactic ambiguity amidst contextual clarity. In *Soft Computing in Humanities and Social Sciences*, ed. R. Seising, and V. Sanz. Springer-Verlag, Berlin Heidelberg.

Carvalho, J.P. 2012. Rule based fuzzy cognitive maps in humanities, social sciences and economics. In *Soft Computing in Humanities and Social Sciences,* ed. R. Seising, and V. Sanz. Springer-Verlag, Berlin Heidelberg.

Chandana, P.H. 2019. A survey on soft computing techniques and applications. *IRJET* 6(4): 1258–1266.

Chang, F., L. Chang, C. Huang, and I. Kao. 2016. Prediction of monthly regional groundwater levels through hybrid soft-computing techniques. *Journal of Hydrology* 541(B): 965–976. Doi: 10.1016/j.jhydrol.2016.08.006.

Das, S.K., A. Kumar, B. Das, and A.P. Burnwal. 2013. On soft computing techniques in various areas. *CS & IT-CSCP*: 59–68. Doi: 10.5121/csit.2013.3206.

Falcone, R., C. Lima, and E. Martinelli. 2020. Soft computing techniques in structural and earthquake engineering: A literature review. *Engineering Structures* 207. Doi: 10.1016/engstruct.2020.110269.

Frankenfield, J. 2020. Artificial Neural Network (ANN). https://www.investopedia.com > terms > a > artificial-neural-netw... (accessed on November 13, 2020).

"Fuzzy Logic," Wikipedia, 2020c. last modified November 13, 2020, https://*en*.wikipedia.org > wiki > Fuzzy_logic.

Gander, L., U. Reffle, C. Ringlstetter, S. Schlarb, K. Schulz, and R. Unterweger. 2012. Facing uncertainty in digitisation. In *Soft Computing in Humanities and Social Sciences,* ed. R. Seising, and V. Sanz. Springer-Verlag, Berlin Heidelberg.

Gen, M., and Y.S. Yun. 2006. Soft computing approach for reliability optimisation: State-of-the-art survey. *Reliability Engineering & System Safety* 91(9): 1008–1026. Doi: 10.1016/j.ress.2005.11.053.

"Genetic Algorithm," Wikipedia, 2020d. last modified November 13, 2020, https://en.wikipedia.org > wiki > Genetic_algorithm.

Ibrahim, D. 2016. An overview of soft computing. *Procedia Computer Science* 102: 34–38. https://doi.org/10.1016/j.procs.2016.09.366.

Jindal, S., and R. Jindal. 2012. Comparing different soft computing techniques for classification of satellite images. *Journal of Information Systems and Communication* 3(1): 52–56.

Ko, M., A. Tiwari, and J. Mehnen. 2010. A review of soft computing applications in supply chain management. *Applied Soft Computing* 10(3): 661–674. Doi: 10.1016/j.asoc.2009.09.004.

Kumar, A., and A. Jaiswal. 2017. Empirical study of twitter and tumblr for sentiment analysis using soft computing techniques. In *Proceedings of the WCECS 1*, San Francisco, CA.
Kumar, A., and N. Sachdeva. 2019. Cyberbullying detection on social multimedia using soft computing techniques: a meta-analysis. *Multimedia Tools Applications* 78: 23973–24010. Doi: 10.1007/s11042-019-7234-z.
Lapresta, J.L.G., and A. Piggins. 2012. Voting on how to vote. In *Soft Computing in Humanities and Social Sciences,* ed. R. Seising, and V. Sanz. Springer-Verlag, Berlin Heidelberg.
"Machine Learning," Wikipedia, 2020e. last modified November 13, 2020, https://en.wikipedia.org > wiki > Machine_learning.
McGraw-Hill Dictionary of Scientific & Technical Terms, 6, s.v. "soft computing," last modified November 3, 2020, https://encyclopedia2.thefreedictionary.com/softcomputing.
Mahdipour, E., and C. Dadkhak. 2014. Automatic fire detection based on soft computing techniques: review from 2000 to 2010. *Artificial Intelligence Review* 42: 895–934.
"Probabilistic Logic," Wikipedia, 2020f. last modified November 13, 2020, https://en.wikipedia.org > wiki > Probabilistic_logic.
Puente, C., A. Sobrino, and J.A. Olivas. 2012. Retrieving crisp and imperfect causal sentences in texts: From single causal sentences to mechanisms. In *Soft Computing in Humanities and Social Sciences,* ed. R. Seising, and V. Sanz. Springer-Verlag, Berlin Heidelberg.
Sanz, V. 2012. How philosophy science and technologies studies, and Frminist studies of technology can be of use for soft computing. In *Soft Computing in Humanities and Social Sciences,* ed. R. Seising, and V. Sanz. Springer-Verlag, Berlin Heidelberg.
Seising, R. 2012. Warren weaver's "science and complexity" revisited. In *Soft Computing in Humanities and Social Sciences,* ed. R. Seising, and V. Sanz. Springer-Verlag, Berlin Heidelberg.
Seising, R., and V. Sanz. 2012. From hard science and computing to soft science and computing – an introductory survey. In *Soft Computing in Humanities and Social Sciences,* ed. R. Seising, and V. Sanz. Springer-Verlag, Berlin Heidelberg.
Termini, S. 2012. On explicandum versus explicatum: A few elementary remarks on the birth of innovative notions in fuzzy set theory (and soft computing). In *Soft Computing in Humanities and Social Sciences,* ed. R. Seising, and V. Sanz. Springer-Verlag, Berlin Heidelberg.
Verikas, A., Z. Kalsyte, M. Bacauskiene, and A. Gelzinis. 2010. Hybrid and ensemble- based soft computing techniques in bankruptcy prediction: a survey. *Soft Computing* 14: 995–1010.
Yildirim, E., A. Gulbag, G. Horasan, and E. Dogan. 2011. Discrimination of quarry blasts and earthquakes in the vicinity of Istanbul using soft computing techniques. *Computer & Geosciences* 37(9): 1209–1217. https://doi.org/10.1016/j.cageo.2010.09.005.
Youth4work. 2020. what is full form of Computer? https://www.youth4work.com > ... > Forum (accessed November 11, 2020).
Yu, W., and Y. Liu. 2006. Hybridization of CBR and numeric soft computing techniques for mining of scarce construction databases. *Automation in Construction* 15(1): 33–46. Doi: 10.1016/j.autcon.2005.01.007.
Zadeh, K.S. 2012. Fuzzy deontics. In *Soft Computing in Humanities and Social Sciences,* ed. R. Seising, and V. Sanz. Springer-Verlag, Berlin Heidelberg.
Zadeh, L.A. 1996. Soft computing and fuzzy logic. In *Advances in Fuzzy Systems – Applications and Theory: Fuzzy Sets, Fuzzy Logic, and Fuzzy Systems,* ed. Klir, and V. Yuan, pp. 796–804. World Scientific Publishing Co., Inc., Hackensack, NJ.
Zadeh, L.A. 1998. Some reflections on soft computing, granular computing and their roles in the conception, design and utilization of information/intelligent systems. *Soft Computing* 2: 23–25. Doi: 10.1007/s005000050030.

12 An Approach Based on Fuzzy Logic for Analysis on Product Development in Open Innovation Context

Ricardo Santos
GOVCOPP - Univ of Aveiro

Antonio Abreu
ISEL, Instituto Politécnico de Lisboa, Portugal/CTS - Nova Univ of Lisbon

J. M. F. Calado
IDMEC/ISEL - Instituto Superior de Engenharia de Lisboa, Instituto Politécnico de Lisboa - Univ of Lisbon

Jose Miguel Soares
ISEG-Lisbon School of Economics & Management, Universidade de Lisboa, Portugal/ ADVANCE, ISEG - Univ of Lisbon

Ana Dias
UNINOVA/ISEL - Instituto Superior de Engenharia de Lisboa, Instituto Politécnico de Lisboa – Univ of Lisbon

Fernanda Mendes
ESAI-Escola Superior de Atividades Imobiliárias

CONTENTS

12.1 INTRODUCTION

Currently, the rapid change of the global market, together with its unpredictability, has increased the competition among enterprises, which has forced some of them to find alternative solutions, to provide a better response to the market changes (Mansor et al. 2016).

Nevertheless, and through the works found on literature, there is (somehow) a consensus that enterprises can achieve sustainable competitive advantage on the existing (or even emergent) markets, by applying innovation on their new products to be launched (Abreu et al. 2018). To do that, some companies have acted in an open innovation (OI) environment, by established collaborative networks to join competences and resources, which otherwise would be more costly if they get it, when acting "alone" and in a closed innovation scenario.

Such collaboration networks are defined in some works (e.g. Abreu et al. 2018; Coras and Tantau 2013; Camarinha-Matos and Afsarmanesh 2007) as virtual enterprises (VE), which is a temporary collaborative network that acts as an organization (Camarinha-Matos and Afsarmanesh 2007). This organization shares different competencies (e.g. training, know-how, patents) and resources as well (e.g. personal, machinery, materials, etc.), in order to develop new products with high level of innovation to provide a better response to the market opportunities and therefore to achieve sustainable competitive advantage (Rosas et al. 2015). However, there are some challenges to be accounted when New Product Development (NPD) takes place on OI context, particularly when it comes to VEs. One of these challenges has to do with several risks, concerning the different activities, considered on the product development process (PDP) as a hole, as shown in Figure 12.1. For each process activity can coexist several risks associated, which can also compromise the hole product, by influencing the risk directly involved with the product itself on its different areas (e.g. sales, finance).

Therefore, and in order to get the benefits from the OI context, it is highly advised from the literature the deployment of suitable management risk models, with the purpose of detecting, prioritizing and reducing the risks involved (Kim and Vonortas 2014).

However, there is an absence of NPD's risk evaluation approaches from the literature that allows to deal with risk and most of all with the possible influence between the product overall risk and the risk resulted from the activities regarding its components (Rossi et al. 2012).

Nevertheless, and by using suitable risk management methods, it is possible to perform a suitable selection of the OI's partner, minimizing, therefore, the risk associated on each product's activity and the product's development risk as well (Kim and Vonortas 2014).

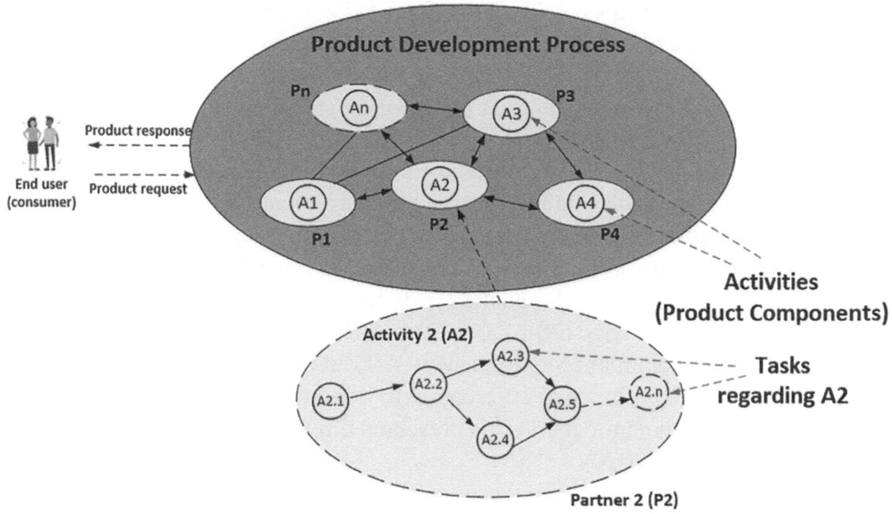

FIGURE 12.1 Process, activities and tasks on a VE.

Each product can be considered as a project where the risks regarding each project activity can be identified, assessed, reported and prioritized, allowing managers to elaborate measures, to minimize the impact, regarding the negative effects that could appear (Rossi et al. 2012).

Besides the absence of risk assessment approaches regarding OI, and particularly VE's, most of the methods existing in the literature don't explore a possible influence among the activities involved with product development, and the risk product itself (Rosas et al. 2015), with most of the methods found, does not include the risk manager perception, on defining the probability of occurrence and also the impact (Abreu et al. 2018; Rossi et al. 2012).

By bearing in mind all of these issues, this work intends to present a risk assessment approach for NPD in OI's context. Thus, and to pursue such purpose, this paper presents the following structure: Section 12.2 describes the literature review, Section 12.3 describes the research method; Section 12.4 the validation by using a real case study, with the results being presented and discussed. Section 12.5 ends with the conclusions and some directions regarding future works.

12.2 LITERATURE REVIEW

Scientific work that approaches the topics of risk analysis, assessment and/or classification, as well as respective models through tools, such as the Failure Mode and Effect Analysis (FMEA), is very scarce and little explored by researchers, linking aspects of open innovation, co-innovation or integrated innovation in collaborative networks. It is permissible that, in a certain type of industry, these issues assume aspects of confidentiality and therefore are not subject to investigation. The work of Lazzarotti et al. (2013) can be an evidence of this where the reference to the risk of the project itself is minimal.

However, the work of Rosas et al. (2015) it is considered paradigmatic, central and decisive, because it addresses central topics, such as: open innovation; risk assessment; collaborative networks and FMEA. And it also presents an approach for risk assessment toward providing an answer to the formulated research question, the benefit from open innovation which is concerned on how to access risk.

Sleefe (2010) presents a risk-based framework that simultaneously optimizes innovation attributes along with traditional project management and system engineering specifications. And more recently, Etges et al. (2017) integrates a model for identifying and managing the degree of risk in companies, using Analytical Hierarchy Process (AHP), Monte Carlo Simulation (MCS) and Balanced Scorecard (BSC) tools. Also, according to Xiao (2016), Case Based Reasoning (CBR) is an analogical AI method that can help managers to identify risks in new technological innovation processes. And the importance of co-creation to minimize risks inherent to develop new concepts inside dynamic markets is presented through an integrated model proposed by Tsutsui et al. (2020).

According to Yang (2010), it is very important to assess the impact of risk on decisions during the development of new products or new systems. And to highlight this importance, this author simulates the behavior and characteristics of products in the future in order to help managers in risk-assessing and decision-making for NPD using the software ARENA by modeling the whole life cycle of a refrigeration system with new features along with new functionalities.

The work of Pereira et al. (2013) aims to lift a framework intended to correct potential biases, caused by different risk attitudes among users and draw conditions to render a concise model to assess project risks within small SMEs. And according to Yun et al. (2020), one of these fields refers to the influence of people in entrepreneurship and intrapreneurship actions as being strong drivers of open innovation promotion. These authors present some models and studies to better identify and understand the behavioral and corporate risks that normally involve NPD processes.

Finally, and with any less emphasis on the issues inherent to risk, but with a focus on the perspective of open innovation, Kim et al. (2020) presented a hybrid algorithm regarding open innovation perspective, where a Vector Autoregressive Model (VAR) is combined with a deep learning algorithm, to predict asthmatic occurrences due to the increment of air pollution, in the field of public health sustainability.

12.3 RESEARCH METHOD

12.3.1 Open Innovation and Virtual Enterprises

Increasingly, the products we use today are developed and produced by VEs, which consist of a given number of collaborative organizations, with the aim of establishing a temporary alliance to act as a unique organization to share resources, skills and competencies, in order to quickly react to the market changes, by conceiving innovative products.

For that propose, a collaborative network (CN) is formed, normally formed by a set of SMEs in order to conceive highly innovative products that better meet the consumer needs, as Figure 12.1 illustrates.

Besides the effectiveness on developing such products, there is the advantage on getting the necessary resources and competencies with a low cost, which was not possible to provide separately. Even if it was, it would be more expensive, which would reduce the efficiency and therefore the competitiveness with the same product. Moreover, CNs can get access to the bigger capital and various technologies due to cooperation with large organizations, which can serve the consumers faster, becoming therefore more flexible.

Therefore, and regarding the OI's context, the NPD can be considered as an integrated process, formed by a set of activities, each one regarding a specific product's/system component, also illustrated in Figure 12.1. Each activity can be performed by one or more OI's partners, involved on behalf of the VE created, and it can be considered as a product/system's component, formed by a set of individual tasks regarding each activity n (An). Each An can be therefore be considered as an individual project that needs to me managed, including its risk. The aggregation of all An's leads to the entire product, which can also be considered an integrated project, required to be managed, given the resources and competences involved.

12.3.2 Project and Risk Management

According to PMI2017, Project Management (PM) consists in the application of skills, techniques, tools and knowledge, in order to conceive activities to attend the project needs.

The use of PM techniques allows organizations to perform projects in an effective and efficient way, by helping organizations to reach a series of benefits, such as the accomplishment of business goals and stakeholders expectations, the management of constraints (e.g. quality, scope, schedule, costs, resources), the optimal use of the enterprises resources as well as the resolution of problems associated to it. Since the goal of each project is to provide value added to an organization or enterprise, the use of PM techniques also allows to manage changes on enterprises, by moving an enterprise from a current state into a desirable one, in order to reach a specific goal, as shown in Figure 12.2.

Based on Figure 12.2, in most organization's projects, especially those related with NPD, the transition from a current state into a desirable one normally involves multiple steps/activities to achieve a specific goal. The PM also allows organizations to define a better answer to the risks that may arise on future, by identifying, improving or even extinguishing failing activities/projects. To do that, the project activities, and processes, are categorized according to the "know-how" areas, as well as the inputs/outputs, the resources and techniques involved. One of the areas, normally involved in PM, is the Project Risk Management (PRM).

The PRM, consists a set of tasks involving risk management, namely the planning of a project, its risk identification and its monitoring and analysis as well, with the aim of determining if a given project is reliable and successful.

According to PMI (2017), the risk categories normally involved in a project are the cost, time, scope and quality, since it is considered the most important factors to achieve success in a given project by many analysts, although in some cases, more risk categories can be included, depending on the propose of the project.

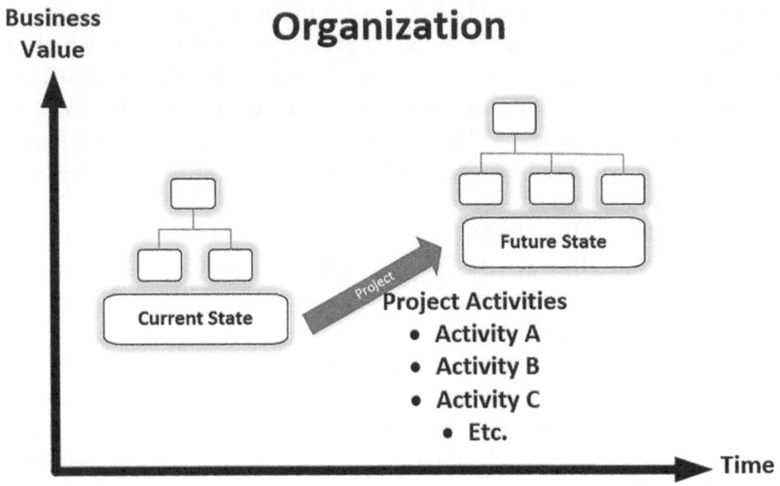

FIGURE 12.2 Organizational State Transition via a Project (adapted from (PMI 2017).

Therefore, and on behalf of PRM, we can also categorize each project activity, according to a set or risks associated.

Based on what was referred before, and regarding OIs, the risk involved in NPD can be assessed by considering each system/component as an individual project, to be analyzed, on behalf of risk management. Since each individual project can bring some risks to the process of NPD, it could be useful to analyze the influence regarding each activity risk n (ARn) over one or more VE's functional areas (FA), by acting as internal factors. The aame principle can be applied for the external factors, which in this case are the ones regarding the external environment of the VE (e.g. political, legal, technological, environmental), that might condition the NPD's success, by influencing its risk. The importance, regarding the risks involved, whether on each ARn, or even in the PR itself, can be defined based on each risk manager perception and by applying methods such as AHP, which shall be described in the next section.

12.3.3 Analytical Hierarchy Process

The AHP method was initially used in 1980 by Thomas Saaty (Ezzabadi et al. 2015), and it allows to help the decision agent to achieve the best solution (based on its preferences) according to a set of conflicting (and also subjective) criteria. By using this method to elicit a set of relative importance indicators (eights), regarding each risk and influence as well, the problem to be solved is then converted into a hierarchical level of elements/criteria.

The importance level is given by the decision-agent, by using a nominal scale, where in this case it is used in the range between 1 and 9, proposed by the author of the method. Through the comparisons made before, as well as their ranks, a matrix is then created, with a set of values being obtained to get the weights, regarding each criterion (Bozbura and Beskese 2007).

12.4 PROPOSED MODEL

The success (or even the threat) of a product to be developed might bring some impact in one (or even more) VE's functional areas (FA) (Etges et al. 2017; Yang 2010). Based on what was referred before, a project can be considered as an entire process, divided into several activities, whose risk can be categorized and analyzed separately by activity, as well together as a whole. Thus, the development of a product can be considered a process, which can be divided into a set of activities. Those activities can be assessed in terms of risk, individually, or even by being linked with the main product (process) itself in order to assess the products' risk, as an internal influence over each risk, regarding each VE's (product) functional domain. The external influence considered here is related with the threats/opportunities, regarding the external environment, divided into several sources of risk (Figure 12.3) (e.g. environmental, technological, economical), as it is found in some works (e.g. Xiao 2016 and Tsutsui et al. 2020).

Therefore, and based on Figures 12.3 and 12.4, each risk activity can be divided into several risks regarding each product component. These activities can even (internally) influence each product's domain, together with the external influence, as can be seen in Figure 12.4.

Additionally, and based on what was referred before, the VE's overall Risk (VER), can be directly achieved through the Product's overall Risk (PR).

According to PMI 2017, the risk involved, regarding a given Activity *n* (ARn) can be achieved based on its several Activities Domains (AD), namely, Quality (Qn), Scope (Sn), Cost (Cn), Tn (Time), Performance (*P*), etc. In this work, we have accounted the following risk categories; Quality (Qn), Cost (Cn) and Time (Tn). According to some studies found in the literature (e.g. Meyer 2015, Etges et al. 2017), there is some influence between the risk regarding the Product (PR) developed on VE

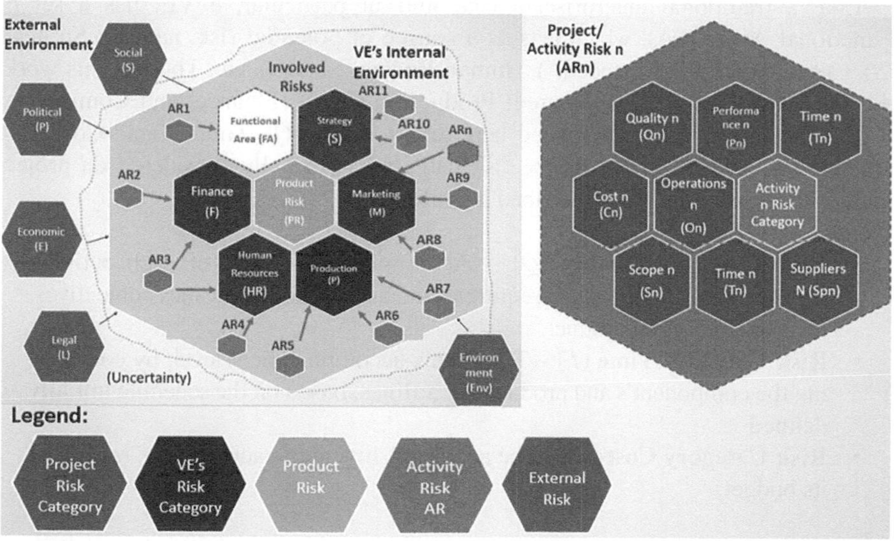

FIGURE 12.3 Risks regarding the new product development in an OI's context.

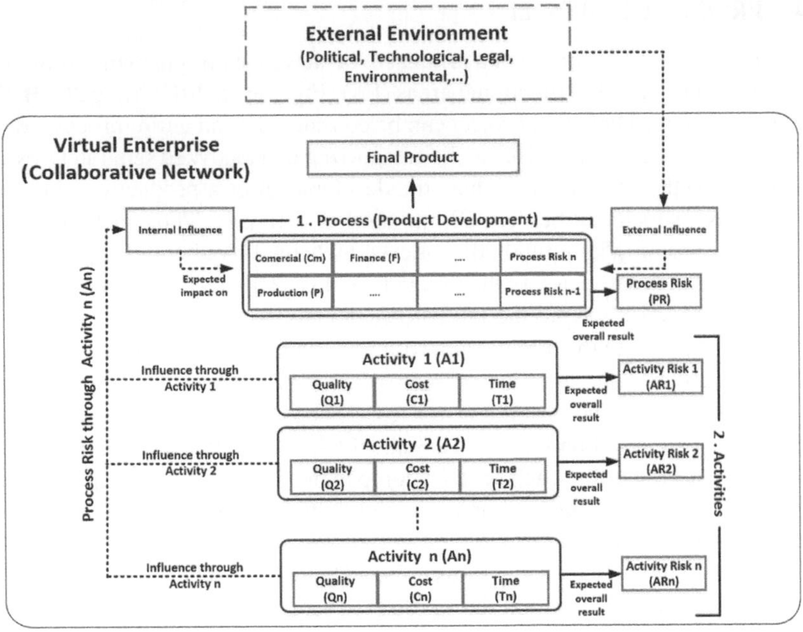

FIGURE 12.4 Relation between the Activity's Risk (ARn) and the Product's Risk (PR).

and the risk regarding each activity (ARn). Such influence cannot only be identified and qualified but also quantified, allowing the prediction of a possible effect regarding a given Activity *n*, over VE's performance, based on the change/transition that occurred between both stages (Etges et al. 2017).

Like a traditional enterprise, a CN, and in particular, a VE, has a set of Functional Areas (FA), which acts as a source of potential risk, namely: Strategy (*S*), Commercial (*C*), Finance (*F*), Human Resources (HR), etc. Thus, in this work, the following FAs were considered: Production (*P*), *F* (Finance) and Commercial (Cm). The risk categories, referred here (e.g. Quality (*Q*), Time (*T*)and Cost (*C*)), are referred on (PMI 2017) as the risk categories normally considered on project management, which can be described as follows:

- **Risk Category Quality (*Q*)** – Accomplishment level of each activity, regarding the component's requirements as well as the business objectives, regarding the final product.
- **Risk Category Time (*T*)** – Timeline's accomplishment level, by considering the component's and product's deadlines, based on the schedule initially defined.
- **Risk Category Cost (*C*)** – The accomplishment of each activity, regarding its budget.

Regarding the FAs of the VE, considered in this work, the following ones were adopted:

- **Production (*P*)** – Concerning the manufacturing and assembly of the final product, as well the problems regarding its supply chain (e.g. problems on product's delivery, delay regarding to the product's time-to-market, delay with some components from a CN's partner, or even from an outside supplier) (PMI2017)
- **Finance (*F*)** – Concerning the product's financing (e.g. credit's exposition to the interest rate, access to private funds), costs involved with the final product (Grace et al. 2014).
- **Commercial (Cm)** – Concerning the hiatus between the perceived satisfaction of the consumer and the product requirements (e.g. the product's technology, the comfort of the vehicle, security) as well as the direct impact by the external environment over its sales (e.g. legal, economic, market competition) (PMI 2017).

Besides the risks considered here, this model can also be extended into other risks. In order to reduce the effects, resulted from the subjectivity degree within human perception on the assessment of risk, a set of fuzzy logic inference systems (FIS), was also included on the same model, in order to achieve an integrated approach, presented on Figure 12.5, where a set of FIS (marked as "*F*") were then used to perform a qualitative (and quantitative) analysis of each risk considered in this model, which was done by integrating the risk concerning the activities and the process as well into a single approach, as shown on Figure 12.5.

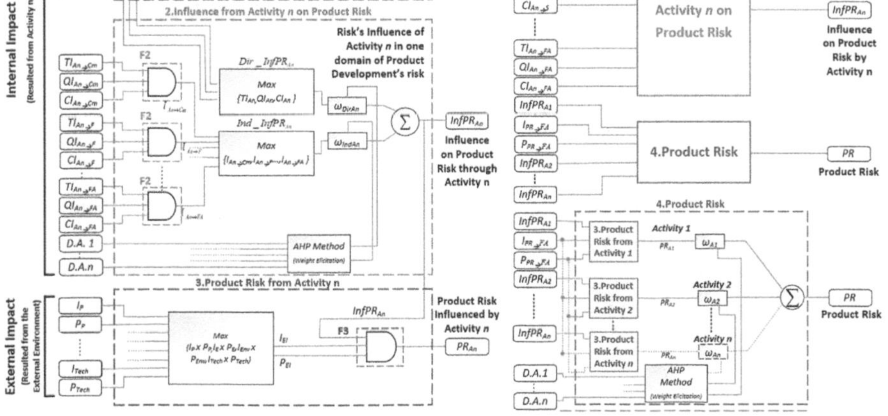

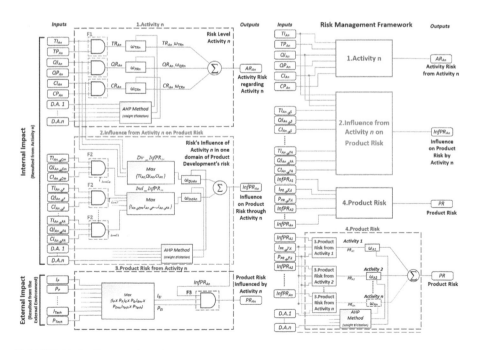

FIGURE 12.5 Proposed model.

Based on Figure 12.5, the Activity Risk Level allows to evaluate the AR_n (Activity Risk related to Activity *n*), while the Product Risk level allows to evaluate the PR (Product Risk) by also accounting the influence regarding each activity *n* on the PR ($InfPR_{An}$).

For each category risk related to each activity considered here (e.g. Time (*T*), Cost (*C*) and Quality (*Q*)), the correspondent risk can be achieved through the associated expected impact and the probability of occurrence as well:

$$TR_{An} = TP_{An} \cdot TI_{An} \tag{12.1}$$

$$QR_{An} = QP_{An} \cdot QI_{An} \tag{12.2}$$

$$CR_{An} = CP_{An} \cdot CI_{An} \tag{12.3}$$

Thus, the Activity Risk related to an activity $n(AR_{An})$ is achieved by weighing all the risks achieved before:

$$AR_{An} = \omega_{TR_{An}} \cdot TR_{An} + \omega_{QR_{An}} \cdot QR_{An} + \omega_{CR_{An}} \cdot CR_{An} \tag{12.4}$$

The weights $\omega_{TR_{An}}$, $\omega_{QR_{An}}$ and $\omega_{CR_{An}}$ are corresponding each one to an AR category and complies the condition:

$$1 = \omega_{TR_{An}} + \omega_{QR_{An}} + \omega_{PR_{An}} \tag{12.5}$$

The existence of some (and possible) influence of each AR_{An} category over each VE's FA, is also accounted here through the corresponding expected impact ($^{-}I_{An\to FA}$).

According to Figure 12.5, and for each process's (or product) FA, only one impact value ($^{-}I_{An\to FA}$) is chosen, which results from the risk manager's perception. Therefore, and for each product's FA, a set of linguistic variables is established followed by a set of inference rules to obtain the respective $^{-}I_{An\to FA}$ values. Based on Figure 12.5, the Ind_InfPR_{An} (indirect influence on product's risk from each AR) is therefore achieved by taking the maximum value from the three correspondent impact values, related to each product's FA ($^{-}I_{An\to FA}$), i.e.:

$$Ind\_Inf_{An} = \max\left\{TI_{An\to Cm}, QI_{An\to Cm}, CI_{An\to Cm}, \ldots, SI_{An\to F}, PI_{An\to F}, CI_{An\to F}\right\} \tag{12.6}$$

On the other hand, the Dir_InfPR_{An} (direct influence on product's risk from each AR), which represents the resources' impact, applied on each activity over the assets deployed on the product's context was also considered here.

Like the Ind_InfPR_{An}, the Dir_InfPR_{An} results from the maximum value of $-I_{An}$, regarding the 3AR's categories defined above:

$$Dir\_InfPR_{An} = \max\left\{TI_{An}, QI_{An}, CI_{An}\right\} \tag{12.7}$$

Which means that the overall influence related to an activity *n* over the process $\left(InfPR_{An}\right)$ is achieved by adding the Ind_InfPR_{An} with Dir_InfPR_{An}:

$$\text{InfPR}_{\text{An}} = \text{Ind\_InfPR}_{\text{An}}.v_{\text{Ind.An}} + \text{Dir\_InfPR}_{\text{An}}.v_{\text{Dir.An}} \tag{12.8}$$

Where ω_{DirAn} and $\omega_{\text{Ind.An}}$, are he weights regarding the direct influence and also the indirect one, expressing therefore, the relevance established by the risk analyzer. The combination of both parameters complies with the following expression:

$$1 = \omega_{\text{Dir.}} + \omega_{\text{Ind.}} \tag{12.9}$$

Therefore, the contribution in terms of risk regarding each activity (ARn) for the total PR, is the result of the occurrence's probability (P_{EI}) combined with correspondent expected impact (I_{EI}) and the overall influence, regarding an activity n over the process (InfPR_{An}):

$$\text{PR}_{\text{An}} = \text{P}_{\text{EI}} \cdot \text{I}_{\text{EI}} \cdot \text{InfAR}_{\text{An}} \tag{12.10}$$

Each *EI* represents the external impact factors over the VE's overall performance and regarding the external environment (Figure 12.5), where according to some literature (e.g. Xiao 2016 and Kim and Vonortas 2014) regarding strategic analysis, such factors can be: Political (*P*); Economical (*E*); Social (*S*); Technological (*T*); Environmental (Env) and Legal (*L*).

In this work there were considered only four, namely:

$$\text{EI} = \left\{\text{Political}\,(P),\ \text{Economical}(E), \text{Environmental}(Env), \text{Technical(Tech)}\right\} \tag{12.11}$$

Thus, the values of P_{EI} and I_{EI} are obtained through the maximum product value between both inputs, concerning the EI's referred on (12.11):

$$\langle I_{\text{EI}}, P_{\text{EI}} \rangle = \max\left\{I_P.P_P, \ldots, I_{\text{Tech}} \cdot P_{\text{Tech}}\right\} \tag{12.12}$$

Therefore, and based on (12.10), the risk regarding the process to develop the product (PR), is the weight sum of each contribution regarding each Activity Risk (PR_{An}), based on its relative importance (ω_{An}) given by the risk manager:

$$\text{PR} = \omega_{A1} \cdot \text{PR}_{A1} + \omega_{A2} \cdot \text{PR}_{A2} + \omega_{\text{An}} \cdot \text{PR}_{\text{An}} \tag{12.13}$$

where ω_{A1}, ω_{A2} and ω_{An} are the weights regarding the activities 1, 2 until n.

All the weights used by the model were achieved by using Analytical Hierarchical Process (AHP) method. To include human perception within the impact and probability's definitions, three types of Fuzzy Inference Systems (FIS) were defined, as illustrated by Figure 12.6.

For the risk levels, concerning TR_{An}, QR_{An} and PR_{An} risks, it was adopted the FIS (marked as "*F*") 1 type, while for the impact level, concerning the $^{-}I_{\text{An}\rightarrow\text{FA}}$, it was used FIS2 type. Finally, and for the PR_{An} risk levels, it was used the FIS3 type (Figure 12.6). For all of the FIS mentioned here, it was defined a set of "if→and

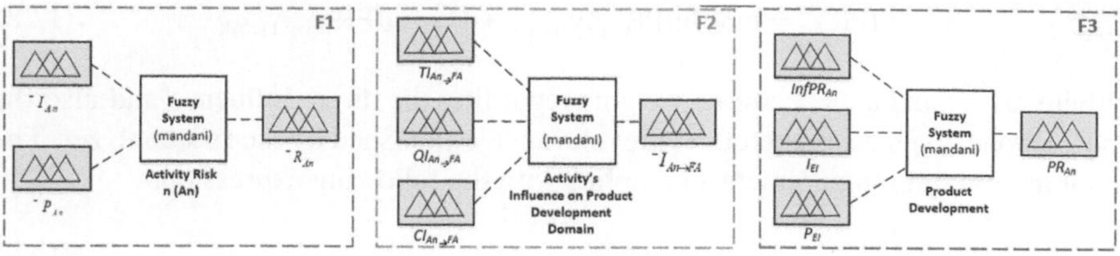

FIGURE 12.6 Fuzzy Inference Systems (FIS) used, regarding the AR, the $InfPR_{An}$ and PR levels.

→then" rules. Based on Figure 12.6, and regarding FIS F1, it was followed the expressions (1–3), according to its variables $^{-}I_{An}$ and $^{-}P_{An}$:

$$^{-}R_{An} = {}^{-}P_{An} \cap {}^{-}I_{An} \tag{12.14}$$

Each rule regrading F1 type can now be defined as "IF $^{-}I_{An}$ is I and $^{-}P_{An}$ is P, then $^{-}R_{An}$ is R". A similar approach was made for FIS2 and FIS3 types, where for the FIS F2, the rules created were based on the following expression, concerning the impact of each AR_{An} category over the *PR* functional areas $\left({}^{-}I_{An\to FA}\right)$:

$$^{-}I_{An\to FA} = TI_{An\to FA} \cap QI_{An\to FA} \cap CI_{An\to FA} \tag{12.15}$$

While for the FIS3, the rules created were based on the following expression, concerning the contribution of an activity An for the Process Risk PR (PR_{An}):

$$PR_{An} = P_{EI} \cap I_{EI} \cap InfPR_{An} \tag{12.16}$$

For each FIS created here, it was considered 5 linguistic levels since that according to some works found on literature (e.g. Ezzabadi et al. 2015), the number of levels should not surpass 9, given the limits with perception related with the decision agent and regarding the levels of discrimination. For each linguistic level, it was defined a triangular pertinence function. Figure 12.7 shows an example of the memberships function regarding the Time (*T*) risk category.

The parameters (*a*, *b*, *c*) regarding the triangular functions are presented on Tables 12.1–12.3, regarding the variable's type of Probability of Occurrence, Impact and Risk.

For both variables used here regarding the probability of occurrence (P_{EI} and $^{-}P_{An}$) it was used the pertinence functions presented on Table 12.1, while the pertinence functions concerning all the expected impacts (e.g. $^{-}I_{An}$, $^{-}I_{An\to FA}$ and I_{EI}), considered in this work, are described on Table 12.2. The pertinence functions concerning all the risks involved (e.g. $^{-}R_{An}$ and PR_{An}) are presented in Table 12.3.

All the FIS's considered, were implemented on Matlab R2016b software, which includes the definition of the inference rules, followed by the membership functions, the fuzzy inference mechanism, the defuzzification method, etc. According to the

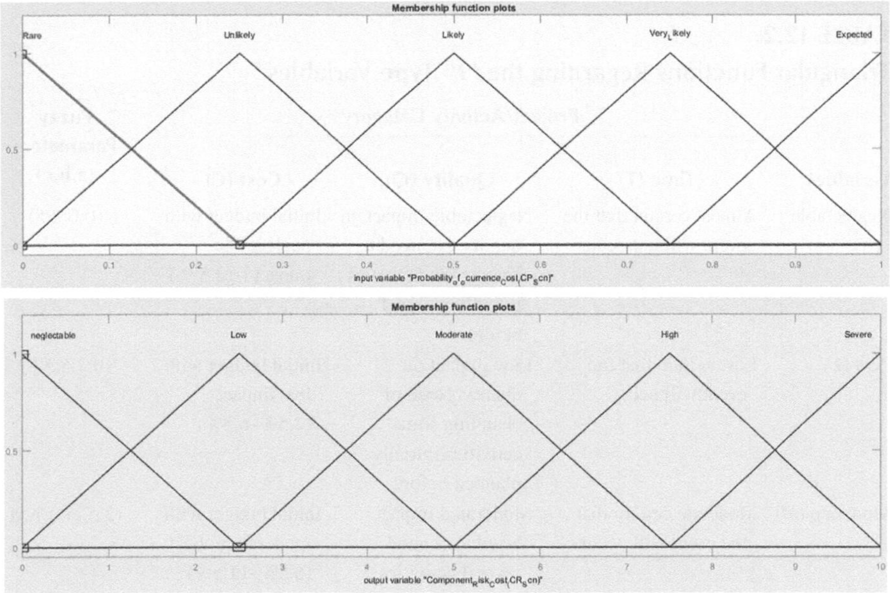

FIGURE 12.7 Membership function related to Time Risk category and "Time" category: (a) TI_{An}. (b) TR_{An}.

TABLE 12.1
Triangular Functions, Regarding the "P" Type Variables

Variables	Detailed Info	Frequency of Occurrence	Fuzzy Parameters (a,b,c)
Rare (*R*)	Event occurrence on exceptional circumstances.	In the next 44 months, the event is expected to occur at least once	(0;0;0.25)
Unlikely (*U*)	Although not likely, the event can occur.	In the next 22 months, the event is expected to occur at least once	(0;0.25;0.50)
Likely (*L*)	The event is probably to be occurs.	In the next 16 months, the event is expected to occur at least once	(0.25;0.50;0.75)
Very Likely (VL)	The likely occurrence of the event.	In the next 10 months, the event is expected to occur at least once	(0.50;0.75;1.0)
Expected (*E*)	Almost sure, that the event occurs.	In the next 4 months, the event is expected to occur at least once	(0.75; 1; 1)

FIS's inputs, the effect of the rules over the output, can be observed from each activity risk (Figure 12.8).

For the defuzzification, the centroid method was used, given its widely acceptance on literature (Abreu et al. 2018) and for the Fuzzy inference mechanism, it was considered the Mamdani inference system given its widely acceptance, as well as its

TABLE 12.2
Triangular Functions Regarding the "*I*" Type Variables

Variables	Project/Activity Category - Time (T) -	- Quality (Q) -	- Cost (C) -	Fuzzy Parameters (a,b,c)
Neglectable (*N*)	Almost certain that the event will not occur	Neglectable impact on quality. No need to change the activities initially planned before	Initial budget with neglectable impact (<2.5 %)	(0;0;2.5)
Low (*L*)	Low certain that the event will occur	Low impact on quality. Need of changing some activities initially planned before	Initial budget with low impact (2.5%–6.5%)	(0;2.5;5.0)
Moderate (*M*)	Moderate certain that the event will occur	Moderated impact. Need to change several activities initially planned before	Initial budget with moderate impact (6.5%–11.5%)	(2.5 ;5.0;7.5)
High (*H*)	Almost certain that the event will occur	High impact. Need of changing almost activities initially planned before	Initial budget with high impact (11.5%–31.5%)	(5.0 ;7.5;10.0)
Severe (*S*)	Practically certain that the event will occur	Significant impact. Need of changing practically all of the activities initially planned before	Initial budget with severe impact (>31.5%)	(7.5 ;10.0;10.0)

TABLE 12.3
Triangular Functions Regarding the "*R*" Type Variables

Level of Risk	Detailed Info	Fuzzy Parameters (a,b,c)
Very low (VL)	Risk acceptance as long as not constituting a threat to the activity/process, although the need of monitoring, to ensure its maintenance level	(0;0;0.25)
Low (*L*)	Although the risk acceptance, its control must be performed based on a cost-benefit perspective	(0;0.25;0.50)
Moderate (*M*)	Need to mitigate the risk involved, which demands the effectiveness on its control and monitorization	(0.25;0.50;0.75)
High (*H*)	The risk should be mitigated as soon as possible, and it should be performed with all the efforts to do it.	(0.50;0.75;1.0)
Very High (VH)	Instant action must be performed to eliminate the risk involved.	(0.75;1.0;1.0)

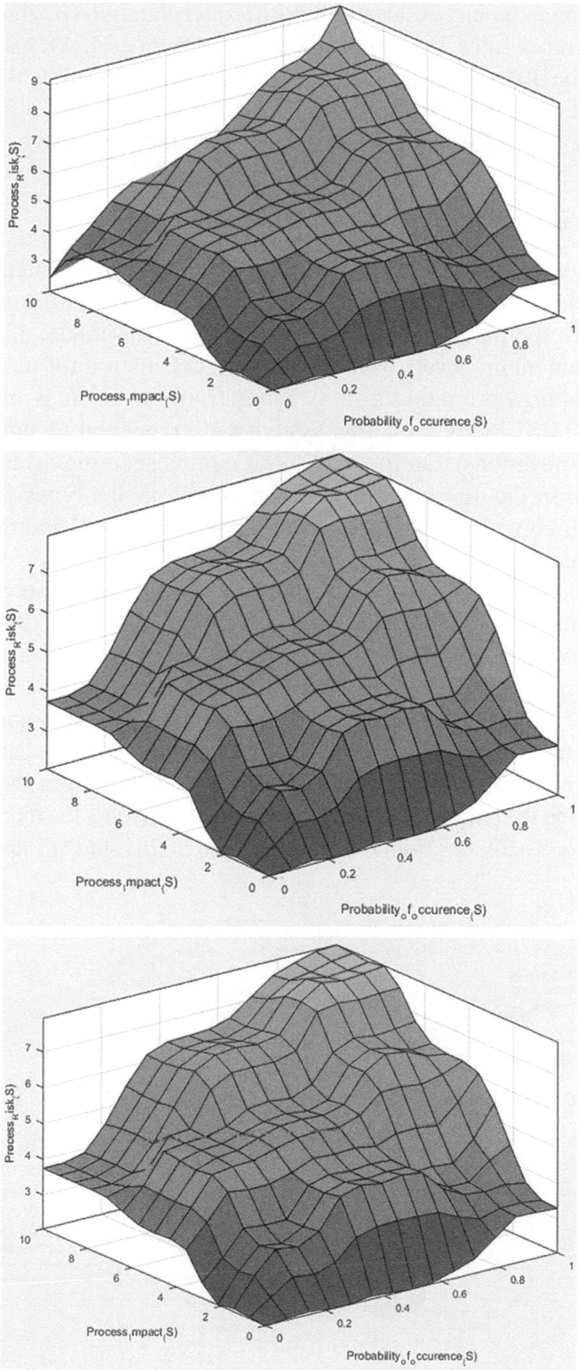

FIGURE 12.8 Rules' surface, from FIS 1 type, regarding each AR considered: (a) Time. (b) Quality. (c) Cost.

suitability to human inputs, besides its intuitive interpretation (Bozbura and Beskese 2007). The inference rules, used to define each FIS from each AR category, explains (in most part) the difference achieved on the three surfaces, respectively correspondent to each AR category. The obtained surfaces also allow to test different combinations, regarding each FIS1's output and for the same input values.

12.5 CASE STUDY

In order to assess the model robustness, a case study was used consisting of the development of an electric vehicle in an OI context. The VE created here was established with the purpose of developing an electric vehicle, aiming to be one of the highest autonomy levels from the market, considered for its class. This was done by developing/optimizing a set of parts from the car (e.g. battery management system (BMS), vehicle traction system with regenerative braking included, among other innovations). On Figure 12.9, it can be seen the diversity of competences involved in the developed of this car as well as the types of entities/organizations also involved here, as well as the partners involved and their interaction on the CN created.

The innovation achieved on the CN composed by 18 organizations/partners was managed by using a model based on the works from Santos et al. (2019).

The CN created here, had the purpose of sharing resources and competencies between the partners/organizations involved to developed and promote a highly innovation product, to better meet the market requirements, raising at the same time the project's credibility from other investors, to gather the right funds for the car's development. Due to the complexity of the project, namely the number of activities and tasks involved, and by considering the purpose of using this case study to assess the model robustness, it was selected only 8 activities from the ones included in this study (Table 12.4).

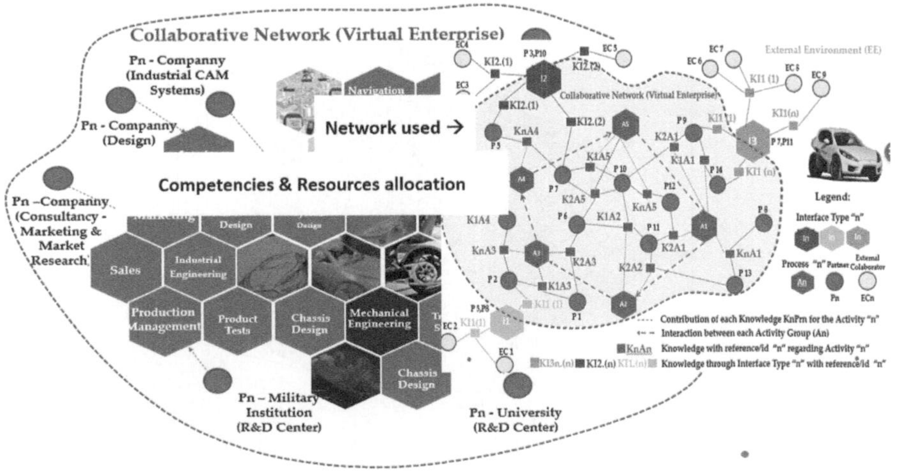

FIGURE 12.9 VE's Collaborative Network (CN): Partners, competencies and resources involved. Source: (Santos et al. 2019).

TABLE 12.4
Selected Tasks to Preform Risk Analysis

Activity nr. (Model's Reference)	VE's Model Reference	Description	Partners
...	...	...	...
1	T2A5	Battery Management System (BMS)	P7,P10
2	T1A4	HMI development software	P4,P3
3	T1A2	Traction control system	P6,P10,P11
...	...	...	...
4	T1A3	Aerodynamic optimization	P2,P1
...	...	...	...
5	T3A5	Chassis design	P10,P12
6	T7A5	Vehicle ESP (Electronic Stability Program)	P6,P7,P10
7	T8A4	Vehicle assembly	P5,P7
8	T7A5	Vehicle driving tests	P6, P7, P10
...	...	...	...

Through Table 12.4, each activity is regarding to a specific part/phase of the process. The responsibility over each partis shared by two (or even more) VE's partners. Each activity is composed by different risks, namely the Quality (e.g. partial/total failure to accomplish the requirements established for the BMS), Cost (e.g. excess cost with the traction control system) and Time (e.g. failure to meet the deadlines with the development of the Vehicle ESP).

12.6 RESULTS AND DISCUSSION

The model's deployment (Figure 12.5) was performed by 8 experts on RM with each one, responsible for the risk management (RM), regarding each activity and representing each one, a different OI's partner. They've assessed each ARn by meeting with the correspondent partners regarding each An and then by meeting only with the others RM's experts that have used the model, in order to assess the product/ process overall risk, by starting to defining the model's inputs, by using FIS F1 to obtain the correspondent outputs (Table 12.5).

The same approach was performed for the FIS F2 and FIS F3. Based on the membership functions, presented in Tables 12.1–12.3, we have converted the linguistic variables, obtained by using FIS F1, F2 and F3, into a quantitative one, by using a suitable scale.

The corresponding results are respectively presented in Tables 12.6–12.9.

Regarding the relative importance factors, represented by the weights used in this model, the AHP method was used, based on the perception of the eight risk managers regarding each An considered, and relatively to each weight used here, namely ω_{CRn}, ω_{TRn}, ω_{QRn} (regarding the ARn level), the ω_{DirAn} and $\omega_{IndirAn}$ (respectively regarding the direct and indirect influence risk level), and the ω_{An} (regarding the contribution of each ARn for the overall Product/Process Risk (PR)) (Figure 12.3). For that purpose,

TABLE 12.5
Model's Inputs/Outputs Regarding the ARn Level

	F1 Inputs						F1 Outputs		
	Time (T)		Quality (Q)		Cost (C)		(T)	(Q)	(C)
An	TI_{An}	TP_{An}	QI_{An}	QP_{An}	CI_{An}	CP_{An}	TR_{An}	QR_{An}	CR_{An}
1	Insignificant	Unlikely	Low	Rare	Low	Unlikely	Low	Very Low	Low
2	Low	Likely	Moderate	Unlikely	Low	Unlikely	Moderate	Moderate	Low
3	Insignificant	Unlikely	Low	Likely	Moderate	Likely	Low	Moderate	Moderate
4	Insignificant	Very Likely	High	Likely	Severe	Unlikely	Moderate	High	High
5	Low	Likely	Moderate	Likely	Moderate	Expected	Moderate	Moderate	High
6	Moderate	Unlikely	Low	Unlikely	Moderate	Unlikely	Moderate	Low	Low
7	High	Expected	Moderate	Rare	Moderate	Unlikely	High	Low	Low
8	Moderate	Likely	Low	Rare	Expected	Likely	Moderate	Low	Moderate

it was collected information by using questionnaires, and regarding the perception of each one of the eight risk managers involved, as well as each CEO's partner, directly involved on each An (eight CEOs on total). The score used here, was the same recommended by (Ezzabadi et al. 2015), which is considered into 10 levels, shown on Figure 12.10.

The importance of using weights here is explained by the fact that all the risks involved (especially those related with the ARn and PR levels) do not have the same importance for the VE's partners and therefore the need to prioritizing them.

Furthermore, the risk prioritization has helped the VE's management board to prioritize the ARn involved, by identifying and analyzing all the risks involved, in order to choose between reducing (or even mitigating) the risk involved or (in alternative) choosing another partner whose behavior can reduce such risk.

TABLE 12.6
Outputs Regarding the Process and Product Risks

	F1 Inputs						F1 Outputs		
	Time (T)		Quality (Q)		Cost (C)		(T)	(Q)	(C)
An	TI_{An}	TP_{An}	QI_{An}	QP_{An}	CI_{An}	CP_{An}	TR_{An}	QR_{An}	CR_{An}
1	1.6	3.2	2.1	1.5	2.1	2.9	2.1	1.9	2.7
2	3.9	5.7	5.1	3.7	2.8	3.7	4.2	4.7	3.2
3	0.7	3.7	3.9	5.3	5.9	4.3	2.9	4.5	5.1
4	1.2	7.5	6.4	5.5	9.8	3.1	5.2	6.4	7.9
5	3.3	5.8	5.8	5.1	4.6	9.1	4.5	5.7	7.7
6	4.7	2.6	2.3	2.7	4.7	3.0	4.3	2.5	3.7
7	6.9	8.3	5.9	0.7	5.2	3.6	7.9	3.5	4.5
8	4.2	5.8	3.7	1.4	9.4	5.6	5.4	3.1	4.9

TABLE 12.7
Outputs Regarding the Process and Product Risks

	F2 (user) Inputs									F2 Outputs		
	Commerce (Cm)			Finance (*F*)			Production (*P*)			(Cm)	(*F*)	(*P*)
An	$TI_{An \to Cm}$	$QI_{An \to Cm}$	$CI_{An \to Cm}$	$TI_{An \to F}$	$QI_{An \to F}$	$CI_{An \to F}$	$TI_{An \to P}$	$QI_{An \to P}$	$CI_{An \to P}$	$I_{An \to Cm}$	$I_{An \to F}$	$I_{An \to P}$
1	0.2	0.8	6.4	0.2	0.2	0.2	2.9	0.2	6.1	2.5	0.2	7.6
2	2.4	4.9	3.9	8.1	5.8	2.4	0.2	9.5	2.4	0.2	8.7	2.9
3	4.6	5.8	4.6	0.2	7.3	0.8	5.4	4.9	4.2	5.8	5.4	4.9
4	6.8	4.2	7.0	6.1	4.2	7.1	2.5	7.1	6.8	2.9	6.9	7.1
5	8.1	9.9	4.0	4.1	2.6	5	3.9	8.7	2.6	2.4	4.3	2.7
6	2.9	5.4	2.5	2.4	6.9	7.7	9.7	5.8	2.9	8.8	4.9	3.2
7	0.2	5.6	4.6	8.7	4.9	2.9	6.7	5.4	7.5	7.3	5.8	0.2
8	1.2	8.6	2.4	0.2	0.2	6.1	0.2	0.2	6.3	0.8	0.7	6.8

TABLE 12.8
Outputs Regarding the Process and Product Risks

	User Inputs								F3 Inputs			F3 Outputs
	Political (P)		Economical (E)		Environment (Env)		Tech (T)		max {IEI×PEI}			
An	I_P	P_P	I_E	P_E	I_{env}	P_{Env}	I_T	P_T	I_{EI}	P_{EI}	$InfPR_{An}$	PR_{An}
1	0.2	3.4	6.8	0.3	0.8	1.5	3.0	1.8	1.7	0.8	5.12	2.7
2	3.8	1.8	3.3	8.1	5.6	3.2	1.2	8.1	3.5	8.1	5.89	6.0
3	0.8	4.7	5.2	5.7	7.9	0.2	4.1	3.7	5.2	5.7	5.86	5.9
4	4.6	2.1	7.2	2.8	5.4	7.1	2.9	6.8	5.4	7.1	8.56	6.4
5	8.1	3.9	2.4	8.1	3.1	8.1	3.5	8.1	8.1	3.9	5.61	5.7
6	3.9	5.7	3.1	5.4	6.8	6.8	8.1	5.6	3.9	5.7	5.33	5.2
7	2.4	7.7	4.9	5.8	5.1	2.5	6.8	4.9	6.8	4.9	7.11	6.9
8	5.8	1.1	2.4	1.4	0.8	6.8	0.2	1.8	5.8	1.1	8.32	5.9

TABLE 12.9
Model's Outputs Related to the Activity Risk (AR) Level

	Time (T)	Qualility (Q)	Cost (C)						
An	TR_{An}	QR_{An}	CR_{An}	ω_{TRn}	ω_{QRn}	ω_{CRn}	$\omega_{Ind.An}$	$\omega_{Dir.An}$	AR_{An}
1	2.1	1.9	2.7	0.29	0.27	0.44	0.55	0.45	2.31
2	4.2	4.7	3.2	0.23	0.24	0.52	0.52	0.48	3.80
3	2.9	4.5	5.1	0.26	0.22	0.52	0.41	0.59	4.39
4	5.2	6.4	7.9	0.37	0.20	0.43	0.46	0.54	6.60
5	4.5	5.7	7.7	0.32	0.15	0.53	0.41	0.59	6.37
6	4.3	2.5	3.7	0.24	0.22	0.54	0.56	0.44	3.58
7	7.9	3.5	4.5	0.28	0.22	0.50	0.52	0.48	5.23
8	5.4	3.1	4.9	0.16	0.27	0.57	0.41	0.59	4.49

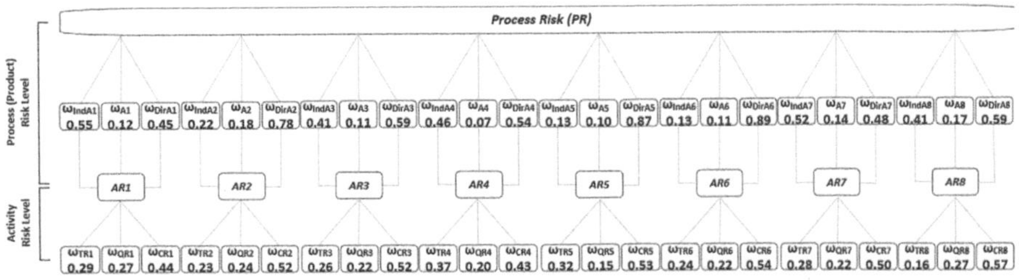

FIGURE 12.10 Model's weight achievement, by using the AHP method.

Based on the results from Tables 12.1–12.3, the overall results regarding the AR level as well as the ones regarding the Product Risk (PR) level (Tables 12.9 and 12.10) were obtained.

Based on the results, shown in Table 12.10, although activities 4 (Aerodynamics Optimization) and 5 (Chassis Design) share the same level of risk ("High"), activity 4 (A4) has the highest risk value, representing therefore the worst threat for the vehicle development. It can also be seen through the same table that both activities are more likely to fail the defined budget, although A4 presents also high risk on Quality (QR_{A4}). Activity A7 presents high risk on meeting the deadlines (TR_{A7}).

From the same table, it is also possible to prioritize each activity according to its overall risk (AR_{An}), where in this case, the An with the worst risk would be defined according to the following sequence: A4, A5, A7, A8, A3, A2, A6 and A1.

On the other hand, and concerning the Product Risk (PR) level (Table 12.10), it can be verified that activities 4 (Aerodynamic optimization) and 7 (Product Assembly) share the same level of contribution to the PR (PR_{A4} andPR_{A7}), although Activity 7 presents the highest value of PR_{An}.

Activities 4 (Aerodynamic Optimization) and 8 (Vehicle Driving Tests) share the same level regarding the highest influence on product risk ($InfPR_{An}$). The activity with more external impact is activity A5 followed by A7, with both sharing the same high impact category (Economy), while the activities more likely to be vulnerable (P_{EI}) to the external impact are A2 (more vulnerable to the technological factor) followed by A4 (more vulnerable to the legal factor, given the vehicle safety requirements settled by the government).

12.7 CONCLUSIONS

This work has presented a method to evaluate the risk regarding the NPD on OI's context, through the VE created to develop the same product. The method presented here also considers the several risks regarding each product/process's activity, as well as a possible influence, concerning each Activity Risk over the Product's Risk through VE's functional areas.

The impact from 10 external environment is also considered here, as well as its likelihood, which allows the risk manager to measure the possible external influence over the Product's Risk, as well as its nature. For the effect of study and the model's robustness, a case study was presented based on the development of an electric vehicle. Based on the obtained results, the development approach allows to prioritize the activities involved, according to its risk, as well as the measurement of the influence degree, regarding each activity risk over the product's overall risk. This last issue is therefore an advantage, when comparing this approach with other methods, because it is also possible to evaluate the risk activity's contribution over PR.

Additionally, and through the activity's prioritization defined before, it can also be known what's the source of risk that most contributes for the highest overall AR_{An}, as well as the partner (s) involved, in order to define the right measures to reduce (or even mitigate) the highest risk (s) detected. Furthermore, the use of fuzzy logic to develop the three types of FIS presented allows to reduce the subjectivity with human perception on risk assessment to evaluate the risks considered in this work.

TABLE 12.10
Model's Outputs Related to Product Risk (PR) Level

	Direct Influence Activity n (Dir.An)			Max	Indirect Influence Activity *n* (Ind.An)			Max	Proj.Infl.	External Impact				
An	TI_{An}	QI_{An}	CI_{An}	$\{TI_{An},\ldots,CI_{An}\}$	$I_{An\to Cm}$	$I_{An\to F}$	$I_{An\to P}$	$\{I_{An}\to Cm,\ldots,I_{An}\to P\}$	$InfPR_{An}$	I_{EI}	P_{EI}	PR_{An}	ω_{An}	$PR_{An}\times\omega_{An}$
1	1.6	2.1	2.1	2.1	2.5	0.2	7.6	7.6	5.12	1.7	0.8	2.7	0.12	0.32
2	3.9	5.1	2.8	5.1	0.2	8.7	2.9	8.7	5.89	3.5	8.1	6.0	0.18	1.08
3	0.7	3.9	5.9	5.9	5.8	5.4	4.9	5.8	5.86	5.2	5.7	5.9	0.11	0.65
4	1.2	6.4	9.8	9.8	2.9	6.9	7.1	7.1	8.56	5.4	7.1	6.4	0.07	0.45
5	3.3	5.8	4.6	5.8	2.4	4.3	2.7	4.3	5.61	8.1	3.9	5.7	0.10	0.57
6	4.7	2.3	4.7	4.7	8.8	4.9	3.2	8.8	5.33	3.9	5.7	5.2	0.11	0.57
7	6.9	5.9	5.2	6.9	7.3	5.8	0.2	7.3	7.11	6.8	4.9	6.9	0.14	0.97
8	4.2	3.7	9.4	9.4	0.8	0.7	6.8	6.8	8.32	5.8	1.1	5.9	0.17	1.00
Process Risk (PR)														5.61

Besides the threats included on risk assessment, the development of this kind of models, particularly the model developed here, should also consider the opportunities that might come within NPD and regarding each activity. The correlation between each activity should also be explored, as well the ways to detect each risk involved.

REFERENCES

Abreu, A., J. M. Martins, and J. M. F. Calado. 2018.A fuzzy reasoning approach to assess innovation risk in ecosystems. *Open Engineering* 8:551–561.

Bozbura, T. F., and A. Beskese. 2007. Prioritization of organizational capital measurement indicators using Fuzzy AHP. *International Journal of Approximate Reasoning* 44:124–147.

Camarinha-Matos, L., and H. Afsarmanesh. 2007. A Comprehensive Modeling Framework for Collaborative Networked Organizations. *Journal of Intelligent Manufacturing* 5:529–542.

Coras, E. L., and A. D. Tantau. 2013. A risk mitigation model in SME's open innovation projects. *Management & Marketing, Economic Publishing House* 8 no. 2 (October):303–328. http://www.managementmarketing.ro/pdf/articole/314.pdf.

Etges, A. P. B., J. S. Souza, and F. J. K. Neto. 2017. Risk management for companies focused on innovation processes. *Production Journal* 27:1–15.

Ezzabadi, J. H., M. D. Saryazdi, and A. Mostafaeipour. 2015. Implementing Fuzzy Logic and AHP into the EFQM model for performance improvement: A case study. *Applied Soft Computing* 36:165–176.

Grace, M. F., T. J. Leverty, R. D. Phillips, and P. Shimpi. 2014. The value of investing in enterprise risk management. *Journal of Risk and Insurance* 82, no. 2 (January):289–316. https://onlinelibrary.wiley.com/doi/epdf/10.1111/jori.12022.

Kim, M.-S., J.-H. Lee, Y.-J. Jang, C.-H. Lee, J.-H. Choi, and T.-E. Sung. 2020. Hybrid deep learning algorithm with open innovation perspective: A prediction model of asthmatic occurrence. *Sustainability* 12, no. 15 (July):1–20. https://www.mdpi.com/2071-1050/12/15/6143.

Kim, Y., and N. S. Vonortas. 2014. Managing risk in the formative years: Evidence from young enterprises in Europe. *Technovation* 8:454–465.

Lazzarotti, V., R. Manzini, L. Pellegrini, and E. Pizzurno. 2013. Open innovation in the automotive industry: Why and How? Evidence from a multiple case study. *International Journal of Technology Intelligence and Planning* 9:37–56.

Mansor, N., S. N. Yahaya, and K. Okazaki. 2016. Risk factors affecting new product development (NPD) performance in small medium enterprises (SMES). *International Journal of Recent Research and Applied Stud*ies 27:18–25.

Meyer, W. G. 2015. Quantifying risk: measuring the invisible. *Paper Presented at the Project Management Institute - PMI® Global Congress EMEA*, England.

Pereira, L., A. Tenera, and J. Wemans. 2013. Insights of individual's risk perception for risk assessment in web-based risk management tools. *Procedia Technology* 9:886–892.

Project Management Institute PMI®. 2017. A guide to the project management body of knowledge PMBOK guide 6th edition. https://www.pmi.org/pmbok-guide-standards/foundational/pmbok.

Rosas, J., P. Macedo, A. Tenera, A. Abreu, and P. Urze. 2015. Risk assessment in open innovation networks. *Paper Presented at the 16th IFIP Working Conference on Virtual Enterprises*, France.

Rossi, B., B. Russo, and G. Succi. 2012. Adoption of free/libre open source software in public organizations: factors of impact. *Information Technology & People* 2:156–187.

Santos, R., A. Abreu, and V. Anes. 2019. Developing a green product-based in an open innovation environment. Case study: Electrical vehicle. *Paper Presented at the 20th Working Conference on Virtual Enterprises PRO-VE*, Italy.

Sleefe, G. E. 2010. Quantification of technology innovation using a risk-based framework. *International Journal of Social, Behavioral, Educational, Economic, Business and Industrial Engineering* 4:868–872.

Tsutsui, Y., N. Yamada, Y. Mitaka, M. Sholihah, and Y. Shimomura. 2020. A strategic design guideline for open business models. *International Journal of Automation Technology* 14:678–689.

Xiao, Q. 2016. Technology innovation risk identification based on sequential CBR. *Paper Presented at the International Conference on Industrial Technology*, Taiwan.

Yang, Z. 2010. Simulation for risk assessment in product development. *Paper Presented at the Second International Conference on Computer Modeling and Simulation*, China.

Yun, J., X. Zhao, K. Jung, and T. Yigitcanlar. 2020. The culture for open innovation dynamics. *Sustainability* 12: no. 12 (June):5076–5097. https://www.mdpi.com/2071-1050/12/12/5076.

Index

Note: **Bold** page numbers refer to tables and *Italic* page numbers refer to figures.

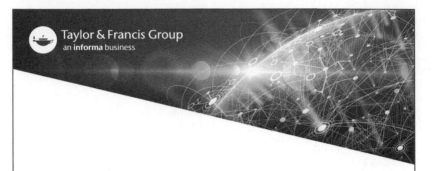